Dave Baum's
Definitive Guide to

LEGO MINDSTORMS

Dave Baum's Definitive Guide to
LEGO MINDSTORMS

Dave Baum

Illustrations by Rodd Zurcher and Dave Baum

Dave Baum's Definitive Guide to LEGO® MINDSTORMS™

ISBN: 1-893115-097

Project Coordinator: Keisha Sherbecoe
Technical Reviewers: Dan Appleman, Idan Beck, and Eddie Herman
Production: TSI Graphics

Distributed to the book trade worldwide by Springer-Verlag New York, Inc.
175 Fifth Avenue, New York, NY 10010
In the United States, phone 1-800-SPRINGER; orders@springer-ny.com
http://www.springer-ny.com

For information on translations, please contact APress directly:
APress, 6400 Hollis Street, Suite 9, Emeryville, CA 94608
Phone: 510/595-3110; Fax: 510/595-3122; info@apress.com; www.apress.com

Printed and bound in the United States of America
1 2 3 4 5 6 7 8 9 10

For Jessica and Jason

Contents

Preface

The MINDSTORMS Robotics Invention System from LEGO is a new kind of toy. True to its heritage, it contains a generous assortment of LEGO pieces that snap, slide, and click into place with amazing simplicity. Nearly all of the pieces can interlock with one another, sometimes in rather unusual ways. What sets MINDSTORMS apart, however, is LEGO's Programmable Brick, called the RCX. Sensors and motors can be attached to the RCX (again with LEGO's hallmark simplicity), and suddenly the RCX brings a LEGO model to life. It not only moves, but also senses and responds to its environment.

Robotics itself is nothing new. Industrial robots have been in use for years and are constantly getting more sophisticated. Likewise, hobbyists have been able to build their own robots for quite a while. Even so, creating robots (even simple ones) has been out of reach for most people. The problem was that constructing robots typically involved soldering, metalworking, and other skills, along with a healthy dose of computer programming. MINDSTORMS has changed all of that. No soldering is required, nor any welding, cutting, or gluing. Even programming has been made easier—a graphical interface allows commands to be "stacked" together almost as easily as LEGO bricks themselves.

MINDSTORMS is an exciting toy. With it, children can learn mechanical design, engineering, and computer programming while they are playing. MINDSTORMS has equal appeal to adults. As a result, MINDSTORMS robots are popping up everywhere—not just in classrooms and toys stores, but also at conventions and, trade shows, and even around the office. This is a good thing. Somewhere along the way many adults forget how important it is to just play, and MINDSTORMS is a nice reminder.

Of course MINDSTORMS only made the process of building a robot easier; designing robots is still a challenging blend of mechanical engineering and computer programming. This can leave many people a little overwhelmed and asking "What do I do now?" That is where this book comes in. Perhaps you're an expert LEGO builder but have never programmed a computer. Or perhaps you just finished writing The World's Most Complicated Computer Program, but you don't know the first thing about gear ratios. Either way you'll find material in this book to guide you through creating MINDSTORMS robots.

The first section covers all of the basics needed to create MINDSTORMS robots. Chapter 4 is especially valuable because it describes a variety of essential construction techniques. Since "hands-on" experience is the most effective form of learning, the second section, on constructing robots, makes up the bulk of the book. In it, step-by-step instructions for building and creating an assortment of robots are provided. Each of the robots demonstrates some interesting construction and/or programming techniques. Together the chapters in this section can almost be considered a course in MINDSTORMS Robotics.

Many of the robots can be built using only the parts contained in the MINDSTORMS Robotics Invention System. However, in several cases some extra parts are required. Often, this is because the purpose of the robot is to demonstrate some feature that can only be achieved with such extra parts. A complete list of extra parts required, including which robots use which ones and where to obtain them, can be found in Appendix B.

You will also need some means of programming the robots. Two different options are presented in the book: RCX Code and NQC. RCX Code is the programming language supported by the official software included with the MINDSTORMS Robotics Invention System. NQC is a free, alternative development system. It is attractive for those who are more comfortable with a traditional approach to programming, require more power than RCX Code provides, or want to program their robots using something other than a computer running Windows. (NQC runs under a variety of operating systems including Windows, MacOS, and Linux.)

The most important thing is to have fun. MINDSTORMS is about playing. There will also be plenty of learning, because you learn through playing, but the key is to play. Experiment with the different robots and change them around. Take things apart and try building them a different way. Change the programming slightly and see what happens. Decorate your robot in your own personal style. Above all, have fun.

Acknowledgements

The instructions for assembling robots are a critical part of the book. A number of options were explored before settling on using 3D renderings. My initial concern about 3D renderings was that creation of accurate 3D models is a difficult and time-consuming business. Actually, the hard part was modeling the individual pieces. Once the pieces themselves were created, putting together an entire robot out of them was quite simple. I would like to thank Rodd Zurcher for creating all of the 3D pieces, as well as taking care of the entire 3D rendering process and serving as a sounding board for all of my ideas. Rodd's attention to detail is what makes the renderings come alive, and I can honestly say that this book would not exist without his enormous contributions.

I would like to thank my wife, Cheryl, for being supportive and understanding throughout this project. Like most things, writing a book turned out to be a lot more work than I first assumed, but Cheryl never once complained about the significant amount of time I was spending on it.

Rodd's wife, Amy, was equally supportive. Thank you for putting up with the late-night calls and visits as Rodd and I sorted through the details of the book.

I would also like to thank Apress founders Gary Cornell and Dan Appleman for giving me the opportunity to create this book. Thank you for your encouragement and patience.

Dan's feedback during the early stages of writing helped establish the style for the entire book. Dan also led the technical review team, which included Idan Beck and Eddie Herman. Their efforts ensured that I presented things appropriately to novices and experts alike.

Thanks also, to Dan Appleman, Idan Beck, and Eddie Herman, my technical reviewers.

Last, I'd like to thank my parents for buying LEGO sets for me in the first place.

PART ONE

Fundamentals

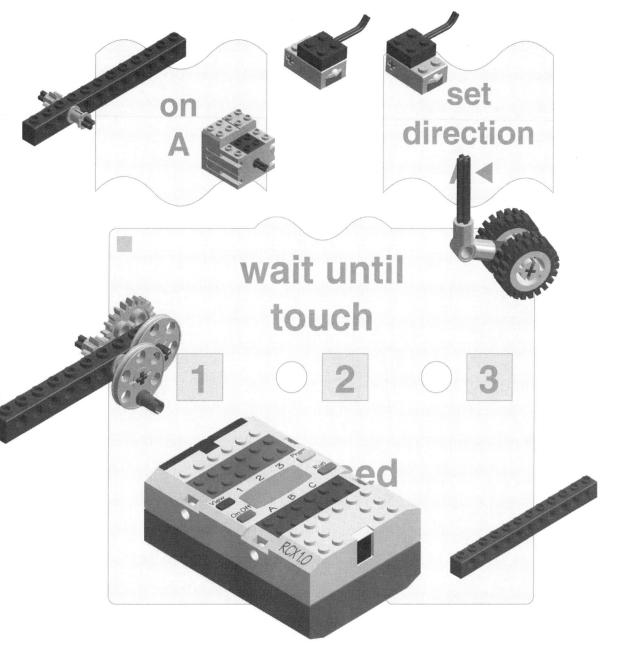

on
A

set
direction

wait until
touch

1   2   3

RCX 1.0

Chapter 1
Getting Started

Welcome to *Dave Baum's Definitive Guide to LEGO MINDSTORMS*. This book represents a journey through the exciting world of LEGO MINDSTORMS. We'll start slowly at first, introducing basic concepts and techniques, then proceed to constructing and programming a number of different robots. By the end of the book, you will be well prepared to build your own robotic creations.

If you haven't yet played with your LEGO MINDSTORMS Robotics Invention System, you should take a few moments now to open it up, read through the *Quick Start* card, and take care of the mundane tasks such as installing batteries and hooking up a serial cable. More detailed information on these tasks can be found in the first few chapters of the *User Guide* supplied with the set.

This chapter introduces the RCX (the brains inside a MIND-STORMS robot) and the programming environments featured in this book. A simple test program will be created, then downloaded and run on the RCX. This test program serves as a convenient way of controlling a motor for some of the examples in subsequent chapters.

This chapter assumes that the infrared (IR) transmitter—included in the LEGO MINDSTORMS set—is already hooked up to your computer, and that batteries are installed in both the IR transmitter and the RCX itself. When communicating, the IR transmitter and the RCX should face one another, about 6″ apart. Communication can be adversely affected by direct sunlight or other bright lights. If communication between the IR transmitter and the RCX is unreliable, try adjusting the position of the RCX and shielding it from unwanted light.

PROGRAMMING ENVIRONMENTS

Programs for the RCX are created on a personal computer (called the *host computer*), then downloaded to the RCX. The RCX can then run these programs on its own without further intervention from the host computer. There are a number of different programming environments that can be used to create RCX programs. The software included with the Robotics Invention System can be used to write programs in *RCX Code*, which is an easy-to-use, graphical programming environment. Although limited in functionality, RCX Code does provide a friendly, intuitive way for beginners to start programming the RCX.

Not Quite C (NQC) is a more traditional, text-based computer language that can also be used to program the RCX. Versions of NQC for Windows, MacOS, and Linux are provided on the accompanying CD-ROM. NQC takes a little longer to learn, but it provides much more power than RCX Code.

RCX Code and NQC are the two languages featured in this book. Sample code is usually provided for both languages, although in some of the more complex cases only an NQC example is possible. Descriptions of some other programming environments can be found in Appendix C.

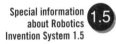

Special information about Robotics Invention System 1.5

The original version of the Robotics Invention System was set #9719. Recently, an updated version called Robotics Invention System 1.5 has been introduced (#9747). There are several differences between these two versions. The 1.5 version comes with updated software, new instructions, and a slightly different mixture of LEGO pieces. Owners of the original version can upgrade to the 1.5 version. See Appendix A for more information.

For the most part, the material in this book applies to both the original and 1.5 version of the Robotics Invention System. There are, however, a few exceptions.

- Two of the projects require pieces from the original set that are not present in the 1.5 set. Further information on obtaining these parts can be found in appendix B.

- The appearance of several of the RCX Code blocks has changed in 1.5. The book uses 1.0-style blocks for illustrating RCX Code programs, so programs written with 1.5 software will look slightly different (although they will behave the same). The most significant changes occur in *my commands*, which are explained in chapter 7.

- Several limitations for RCX Code programs have been removed in 1.5. Most notably, *stack controllers* may be nested and *my commands* can call one another. This is explained further in chapter 7.

- The RCX in the 1.5 set cannot be powered from an external AC adapter.

The next few sections will guide you through the installation and use of NQC. If you wish to use RCX Code instead, you can skip to the section titled *RCX Code Quick Start* later in this chapter.

When the RCX is powered up for the first time, a special piece of software called *firmware* must be downloaded to the RCX. This firmware provides a sort of operating system for running your own programs. If you write programs in RCX Code, the firmware will automatically be downloaded whenever it is needed. If you use NQC, you must download the firmware manually. If the RCX's display looks like the illustration below, firmware has not yet been downloaded. Once it is loaded, the RCX will remember the firmware even when it is turned off.

Figure 1-1
Firmware-Needed Display

The firmware itself is on the Robotics Invention System CD-ROM (not this book's CD-ROM) in a folder named **FIRM**. The name of the file depends on its version number. As of this writing, the current firmware file is **FIRM0309.LGO**; however, it is conceivable that a newer firmware version will be on future CDs. Check your own CD or use the latest file in the **FIRM** directory. There are actually several different NQC programming tools, and each of them has different ways of downloading firmware. Details are provided in the *Using RcxCC, Using MacNQC,* and *Using NQC for Windows sections* below.

NQC QUICK START

NQC is a textual computer language. Programs are written in an editor, then compiled and downloaded to the RCX. The NQC compiler is primarily available as a command line tool. This means that it must be invoked by typing the proper commands into a *command shell* (such as the MS-DOS Prompt for Windows 95/98). Integrated Development Environments (IDEs), which provide a program editor and a graphical user interface to the NQC compiler, exist on some platforms. RcxCC is one such IDE that runs under Microsoft Windows, and MacNQC is a Macintosh-based IDE (both are included on the accompanying CD-ROM). IDEs make using NQC a bit easier and more friendly, and they are, in general, the preferred solution where available. One important point, however, is that all of these solutions use the exact same language to specify an NQC program. For example, a program written with RcxCC can also be used with the Linux version of the NQC compiler. Quick Start information is provided for three different versions of NQC: RcxCC, MacNQC, and NQC for Windows.

Using RcxCC

RcxCC relies on one of the software components normally installed by the RCX Code software. For this reason, it is advisable to install the RCX Code software prior to using RcxCC. Refer to the *Robotics Invention System User Guide* for more information on installing this software.

RcxCC has its own folder within the **Tools** folder of the CD-ROM. To install RcxCC, run the **Setup.exe** program found within the **RcxCC** folder. After installation, an **RCX Command Center** folder will be added within the **Start** menu's **Programs** folder.

When you launch RcxCC it will display a dialog box that allows you to specify which serial port the IR transmitter is connected to. You can either select the appropriate COM port or let RcxCC automatically check each port and decide for itself. The dialog box also allows you to choose between **Mindstorms** and **CyberMaster**. Assuming that you are using an RCX, you may leave this item set to **Mindstorms**. Make sure the RCX is turned on and facing the IR transmitter, then click the **OK** button.

If RcxCC has trouble communicating with the RCX it will display an error and let you try again. If you click **Cancel**, RcxCC will still start, but those functions that require communication with the RCX (such as downloading a program) will be disabled.

Once started, RcxCC will show two windows: the main window and a floating **templates** window. The **templates** window is a sort of cheat sheet for remembering the various NQC commands. For our present example we can ignore it.

If firmware has not been previously downloaded to the RCX, it should be downloaded now. Select menu item **Tools > Download Firmware** and choose the firmware file to download (it should be on the MINDSTORMS CD-ROM). A progress bar will appear while the download is in progress (it will take a few minutes). Once this is completed, the RCX will be ready to use.

Select menu item **File** > **New** in the main window to create a new NQC program. Inside the program's window enter the following text:

```
task main()
{
}
```

This is an extremely simple NQC program. It does nothing. That won't stop us from compiling and downloading it to the RCX, though. Select menu item **Compile** > **Download**, which will compile the program and download it to the RCX. (If RcxCC was not able to communicate with the RCX at start-up, then **Download** will be disabled.) If any errors were encountered during compiling, you may click on the error messages to see the offending program line. If no errors were present, a sound from the RCX will confirm that the program was indeed downloaded.

That's all there is to editing, compiling, and downloading NQC programs with RcxCC. You can now skip to the section titled *The START Program.*

Using MacNQC

A copy of MacNQC can be found in the **MacNQC** folder within the **Tools** folder on the CD-ROM. To install MacNQC, simply drag the MacNQC application from this folder to your hard drive.

If you need to download firmware to the RCX, select **Download Firmware** from the **RCX** menu. You will be prompted to choose a firmware file, which will then be downloaded to the RCX.

When you launch MacNQC it will create an untitled window in which you can enter a new program. Inside this window type the following text:

```
task main()
{
}
```

It isn't much of a program, but it will at least allow you to familiarize yourself with the process of creating an NQC program. To compile and download this program to the RCX, select

Download from the **RCX** menu. If you made a mistake typing in the program, one or more errors will be shown: otherwise MacNQC will attempt to download the compiled program to the RCX. The RCX will play a sound to confirm that the program was successfully downloaded. MacNQC assumes that the IR tower is connected to the modem port. If you are using a different serial port, select **Preferences** from the **Edit** menu and choose the appropriate serial port in the pop-up menu.

That's all there is to editing, compiling, and downloading NQC programs with MacNQC. You can now skip to the section titled *The START Program*.

Using NQC for Windows

In order to use NQC under Windows, the NQC command (nqc.exe) must be put in a location where the command shell can find it. One solution is to create a single folder containing the NQC command and the programs you create. Copy the file **nqc.exe** from the CD-ROM to such a folder on your hard disk. The remaining instructions assume that you copied the file to a folder named NQC on disk C:, although any folder and drive will do.

Under Windows 95/98, start a command shell by selecting **Programs > MS-DOS Prompt** under the **Start** menu. This will create a window in which you may type commands. First we need to tell the command shell to work inside the NQC folder. The folder that the command shell is working in is called the *current directory*. Type the following command and hit the <return> key to instruct the command shell to use the NQC folder as the current director:

```
cd c:\NQC
```

If you see an error message, then double-check the spelling and location of the NQC folder that was created on your hard disk. If firmware has not been previously downloaded to the RCX, you will need to download it now. The NQC command has a special option for doing this, but you must tell it where to find the firmware file. The firmware can be found in the FIRM directory on your Robotics Invention System CD-ROM. Assuming that the IR transmitter is connected to the first serial port (COM1), turn on the RCX and type the following command (substituting the appropriate drive letter in place of E: for your CD-ROM and the appropriate firmware name, if different):

```
nqc -firmware E:\FIRM\FIRM0309.LGO
```

The download will take several minutes to finish, but once it is completed, the RCX will be ready to use.

By default, the NQC command assumes that the IR transmitter is connected to the first serial port, COM1. If a different serial port is used, then any NQC commands should start with an option of the form "-*Sport*," where *port* is the name of the serial port. For example, downloading the firmware using the second serial port (COM 2) would look like this:

```
nqc -SCOM2 -firmware E:\FIRM\FIRM0309.LGO
```

To create an NQC program we need to use a text editing program. Windows comes with a very simple editor called *Notepad*. Use the following command to create a program called **trivial.nqc**.

```
notepad trivial.nqc
```

Since **trivial.nqc** doesn't exist yet, Notepad will ask if you want to create the file; click **Yes** in the dialog box. In Notepad's window, type the following text:

```
task main()
{
}
```

After entering the text, select **File > Save** in Notepad's menus. Then click on the command shell window and type the following command:

```
nqc -d trivial.nqc
```

This will invoke the NQC compiler, instructing it to compile the program you just created and download it to the RCX. If NQC prints any error messages, check the **trivial.nqc** program and make sure it reads exactly like the listing shown above. If you need to change the program, simply edit it in the Notepad window, then save the changes and compile again. To compile and download a different program, replace **trivial.nqc** with the other program's name.

As mentioned before, **nqc.exe** must be put in a place where the command shell can find it. Our current solution is to keep **nqc.exe** in the same folder as the programs we are writing (such as **trivial.nqc**). Another option is to copy **nqc.exe** to some other folder that is already in the *command path*. The command path is a list of directories that the command shell searches when looking for commands. You can display the command path by typing the following command:

```
PATH
```

This will print a list of directories separated by semicolons. For example, on my computer I see the following:

```
C:\WINDOWS;C:\WINDOWS\COMMAND
```

This means that two directories are searched: C:\WINDOWS and C:\WINDOWS\COMMAND. If I copy **nqc.exe** to either of these two directories, the command shell will be able to find it regardless of what the current directory is set to. Another option would be to add C:\NQC to the search path.

RCX CODE QUICK START

If you plan to use RCX Code, you should follow the instructions provided in the *Robotics Invention System User Guide* for installing and running the necessary software. The installation process is straightforward, and the instructions are clear, so I won't repeat them here.

THE START PROGRAM

Our program will be quite simple; the goal is to start a motor running, then flip the motor's direction each time a touch sensor is pressed. Before writing the program, however, we need to hook up the touch sensor and motor to the RCX, as shown in Figure 1-2.

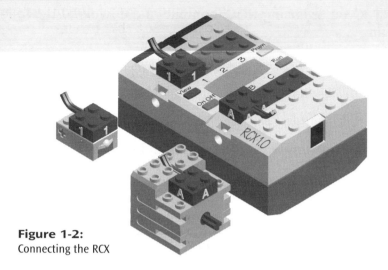

Figure 1-2:
Connecting the RCX

There is something odd about this picture—all of the wires have been cut! For such a simple illustration it would have been possible to show the actual wires, but as the number of motors and sensors increases, these wires become a rat's nest and it is impossible to tell which wire goes where. To make the illustrations clearer, each end of the wire is labeled with either a number or a letter indicating the input or output port on the RCX that the wire will be connected to. For example, the wire attached to the motor is labeled "A," indicating that it should connect to output A. (The labels on the ends attached to the RCX are somewhat redundant.)

Now we're ready to write the program. Don't worry if you don't understand it—explanations will follow later. For now, we are just concentrating on how to use the RCX itself. If you are using RCX Code, create and download the program shown in Figure 1-3.

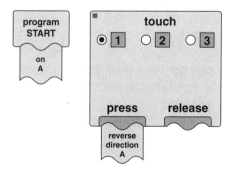

Figure 1-3:
Getting Started Program in RCX Code

If you are using NQC, compile and download the following program:

```
// start.nqc - a very simple program for the RCX

task main()
{
    SetSensor(SENSOR_1, SENSOR_TOUCH);
    On(OUT_A);

    while(true)
    {
        until(SENSOR_1 == 1);
        Toggle(OUT_A);
        until(SENSOR_1 == 0);
    }
}
```

Once the program is downloaded, you can start it by pressing the **Run** button. The LCD should show an animated rendition of a running person, and the motor should start spinning. If this is not happening, check the program and your connections.

If you look carefully, you should also see a small triangle in the LCD above the **A** output port. There are two such indicators for each port, one for the forward direction and one for reverse. This allows you to easily determine which motors are currently running (and in which direction).

If you press the touch sensor, a small triangle will appear on the LCD display just below the sensor's connector. This triangle indicates that the sensor has been activated. Although these indicators are displayed regardless of the type of sensor being used, it is most meaningful with a touch sensor. Pressing the touch sensor will also cause our program to respond—specifically, it will reverse the motor's direction. You can continue to press the touch sensor; each time it will flip the motor's direction either from forward to reverse or from reverse to forward.

The program can be stopped by pressing the **Run** button. Stopping a program in this way will also automatically stop any motors that are running.

THE VIEW BUTTON

By default, the RCX normally displays the value of its internal clock, called the *watch*. Time is displayed in hours and minutes and is reset to 00.00 whenever the RCX is turned on. The RCX is also able to display the current reading from a sensor or the status of a motor. The View button is used to cycle through these various display modes. When a sensor or output port is being viewed, a small arrow appears on the LCD pointing to the port that is being monitored.

Pressing the **View** button once will switch from displaying the watch to displaying sensor 1. Assuming that the touch sensor is not currently being pressed, the display will read 0. Pressing the touch sensor will change the displayed value to 1.

Pressing **View** three more times will display the status of output **A**. Assuming that the motor isn't presently running, the status will be 0. If you run the program (hence starting the motor), the display will then show a value of 8. The RCX can run the motors at 8 different power levels, and the number 8 indicates that the output is at full power.

The **View** button can also be used in conjunction with the other buttons to control a motor. If the **View** button is held down and the display is presently showing an output's status, the **Run** button will turn the output on in the forward direction and the **Prgm** button will turn the output on in reverse. The output will remain on only while the **Run** or **Prgm** button is being held down.

Let's start at the default display mode by turning the RCX off, then back on. The display should be showing the clock (00.00), and the motor should be stopped. Press the **View** button 4 times, but do not release it on the fourth press. While still holding the **View** button down, try pressing the **Run** or **Prgm** buttons to activate the motor.

ATTACHING MOTORS

Wires are attached to the RCX using a special square connector. There are four ways that a wire may be attached to such a connector (the wire may come from the front, right, back, or left). For sensors, the orientation of the connector makes no difference whatsoever. For motors, it determines in which direction the motor spins (clockwise or counterclockwise) when the RCX activates the output in the forward direction.

In our example, the wire extends down from the RCX output port and sits within the groove along the top of the motor. In this orientation, the motor will spin clockwise when the output is activated in the forward direction. You can verify this by running the sample program again. If you have trouble telling which way the motor is spinning, attach a large wheel to it. Try changing the orientation of the connectors on the RCX or the motor and observe how it is possible to make the motor spin counterclockwise.

When building a robot, it is often helpful to change the orientation of the wires so that the output's forward direction corresponds to some obvious motion for the robot itself (moving a vehicle forward, raising an arm, etc.).

DEBUGGING

Sometimes a robot doesn't behave as expected. Isolating the cause of the problem and solving it is called *debugging*. Some common items to check during the process of debugging are presented below:

- Verify that all motors and sensors are correctly wired to the RCX. For motors, not only must the wire attach to the correct output port, but the orientation of the wire at both ends must also be correct; otherwise the motor's rotation may be reversed.

- Use the **View** button to monitor a sensor's value as the program is being run. Verify that the actual sensor values match what the program expects to see. For example, when using a light sensor, be sure that the readings are consistent with any thresholds set in the program. When using touch sensors, verify that they are being pressed and released as expected; sometimes the sensors do not get pushed in firmly enough and will not have the expected value.

- Use the **View**, **Run**, and **Prgm** buttons to manually activate each of the robot's motors. Verify that the motor turns in the correct direction and that any mechanisms driven by the motor are operating properly.

- For RCX Code programs, verify that the correct sensor number is selected within the various program blocks. In some cases the RCX Code software will automatically pick the correct sensor, but sometimes it guesses wrong. When using touch sensors, it is also important to make sure that the code blocks have the correct icon for the button being either pressed or released.

MOVING ON

Now that you know how to write and download a program to the RCX, we can proceed to more interesting material. Before moving on, however, I'd like to leave you with a few pieces of advice.

All of the programs shown in the book are also available on the accompanying CD-ROM (in a folder named **Examples**). In many cases you will find it easier (and less error-prone) to use the programs off the CD-ROM, rather than reproducing them manually.

The RCX tends to have a voracious appetite for batteries, so the use of rechargeable batteries is strongly recommended. In general, rechargeable alkaline cells (such as Rayovac Renewal batteries) perform much better than either NiCad or NiMH rechargeables.

Another battery-saving measure is to run the robots on a smooth surface such as wood or tile, rather than on carpeting. Motors must work much harder to propel a vehicle across carpet, resulting in slow-moving robots that consume batteries rapidly.

The MINDSTORMS Robotics Invention System contains a large number of pieces, including many small ones with very special purposes. Finding the correct piece can become a tedious task and distract from the overall process of building and experimenting with a robot. Thus, I find it helpful to keep the pieces sorted and to store them in small, multicompartment plastic containers such as those used to store assorted hardware or fishing tackle.

Now, on to the next chapter, which explains what the RCX is and what it can do.

Chapter 2
The RCX

At the heart of every LEGO MINDSTORMS robot is the LEGO Programmable Brick—the RCX. Various sensors and motors may be connected to the RCX, allowing it to perceive and interact with the world. There are several different ways to program the RCX, but they all share a set of common functions—namely those of the RCX itself. This chapter describes how the RCX works and what it can do.

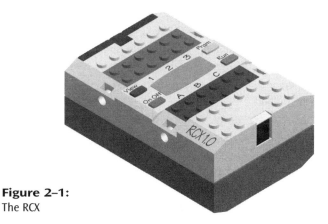

Figure 2–1:
The RCX

THE HARDWARE

The RCX is actually a tiny computer based on a Hitachi H8 series microprocessor. This 8-bit CPU provides most of the control logic for the RCX including serial I/O, Analog to Digital conversion, and built-in timers. It even contains 16K of internal ROM which is pre-programmed with some low-level operating system software.

The RCX also contains 32K of static RAM. Most of this memory is occupied by firmware (discussed below) and various system parameters. However, 6K is reserved for "user memory"; this is where the programs you write for the RCX reside. Compared to a desktop computer, 6K sounds too small to be of any practical value. However, within the RCX, programs tend to be only hundreds of bytes long (rather than the millions of bytes for desktop applications); thus 6K is more than adequate.

The RCX also contains special circuitry to interface with the real world. An LCD and four pushbuttons are provided for user interaction. Special driver chips allow the RCX to control motors or other electric devices attached to the output ports, and the internal ADC (Analog to Digital Converter) allows the RCX to read its three sensor ports.

The RCX uses IR (infrared light) to communicate with a desktop computer or another RCX. IR communication for the desktop computer is provided by an IR interface (included in the Robotics Invention System), which attaches to a standard 9-pin-serial port. Many laptop computers (and palmtop devices) already have their own IR port. All IR devices use infrared light for communication, but the protocols they use to communicate can differ widely. Most laptops and other computing devices use the IrDA protocol for communication, while the RCX uses a simpler proprietary protocol. Because of this, a laptop's IR port will not automatically be able to communicate with the RCX. Of course, anything is possible with the right amount of hacking, but as of this writing I know of no general-purpose solution to making a laptop IR port communicate with the RCX.

The RCX can be powered by 6 AA batteries (some versions can also be powered by an optional AC adapter). How long the batteries last is highly dependent on how the RCX is being used. The biggest drain on the batteries comes from the motors that the RCX must power. When running, motors can consume considerable power, and this power increases when the motor is under great strain. Needless to say, heavy use of the RCX will result in lots of battery changes—using rechargeable alkaline batteries is a good idea. The firmware and any programs that have been downloaded are remembered even when the RCX is turned off. If the batteries are removed for any significant amount of time, however, this memory will be erased. When you need to change batteries, make sure the RCX is turned off, then quickly remove the old batteries and insert new batteries. As long as you work reasonably fast, the memory will be retained and you will not need to reload the firmware.

Another option would be to connect the AC adapter (if available) while changing the batteries.

FIRMWARE

The RCX has its own operating system, which is split into two parts. The first part is stored in ROM and is always present. The second part is stored in RAM and must be downloaded to the RCX the first time the RCX is turned on. This second part is often called the RCX *firmware,* although technically both parts can be considered firmware. This operating system performs all of the mundane functions such as keeping track of time, controlling the motor outputs, and monitoring the sensors.

> Firmware is a generic term for software that is built into a device. The term is intended to convey the fact that such software is something in between *hardware* and the common notion of *software.*

When a program is written for the RCX, it does not contain native code for the CPU. Instead, it consists of special bytecodes which are interpreted by the firmware. This is similar to the way Java programs are stored as Java bytecodes which may be then interpreted by a Java Virtual Machine. Using bytecodes, rather than native machine code, allows the RCX to maintain a safe and reliable execution environment for programs. Although it is possible to write and load custom firmware that would completely take over the RCX hardware, most RCX programming is done within the confines of the standard firmware (hereafter referred to as simply the firmware).

Although it is possible to alter the RCX's capabilities with custom firmware, this book assumes that the standard firmware is being used. Hence, when the RCX is stated to provide certain functions what is really meant is that the "RCX with standard firmware" provides those functions.

TASKS AND SUBROUTINES

An RCX program may contain up to ten *tasks.* Each task is a sequence of bytecodes, which is simply a list of instructions to be followed. Tasks are either *active* or *inactive.* When the task is active, the RCX executes the instructions listed for the task. When a task becomes inactive, the RCX no longer executes its instructions. When a program is started, its first task is made active, and any other tasks are inactive. Stopping a program is equivalent to making all of its tasks inactive.

Multiple tasks may be active at the same time, and in this case the RCX switches back and forth between the active tasks, executing a little bit of each one, so as to make it appear that all of the active tasks are running at the same time. This is known as *concurrent execution,* which has both advantages and disadvantages. The advantage is that for functions that are relatively independent, placing them in separate tasks often makes their instructions much simpler. The disadvantage is that if the separate functions need to interact with each other, concurrent execution can lead to some very subtle bugs (see chapter 7 for an example of this).

A program may also define up to 8 subroutines, which are also sequences of instructions to be executed. The difference between a task and a subroutine is that tasks all run concurrently with one another. Subroutines do not run by themselves; they must be called from a task. When called, the subroutine is executed, but the task that called it must wait for the subroutine to complete before continuing with its own instructions. Any task may call any subroutine, but a subroutine may not call itself or another subroutine. Such restrictions limit the usefulness of subroutines in many applications.

OUTPUT PORTS

The RCX has three output ports (**A, B,** and **C**), each of which can be in one of three modes: *on*, *off*, or *floating*. The *on* mode is just what it sounds like—any motor attached to the output will be running. In the *off* mode, the RCX turns off the output and the motor will be forced to stop. The *floating* mode is somewhat unusual. In this mode, the RCX is no longer powering the output, but a motor is still allowed to spin freely. In some cases, this will have a much different effect than *off*. In terms of a car, *off* is like applying the brakes, while *floating* is more like coasting in neutral.

Each output also has a direction associated with it: either *forward* or *reverse*. This direction only has effect when the output is on, but the setting is remembered and can even be modified while the output is turned off (or floating). As mentioned previously, the actual direction of a motor's rotation (clockwise or counterclockwise) depends upon how the wires are attached between it and the RCX.

The *power level* of an output may also be adjusted to one of eight settings. Like the direction setting, the power level only has effect when the output is turned on, but is remembered and may be altered when the output is turned off. The RCX is primarily a digital device, and digital devices like things to be either on or off, and not be "half on" or "three-quarters on." Hence the RCX needs some way to create these intermediate power levels from a digital signal.

One way of doing this is with Pulse Width Modulation (PWM). Instead of turning a signal on and leaving it on, PWM rapidly switches back and forth between on and off. The amount of time that is spent "on" is called a pulse, and the duration of this pulse is called its width. The percentage of time that the signal is "on" is called its duty-cycle. When using PWM, intermediate levels of power are created by varying the pulse width to generate the appropriate duty-cycle.

In the case of the RCX, the pulses are sent every 8ms. At the lowest power level, the pulse is 1ms long; thus power is supplied only 1/8 of the time (duty-cycle = 12.5%). Higher power levels result in longer pulses, until at the highest power level the pulse actually takes the entire 8ms; in this case power is supplied continuously.

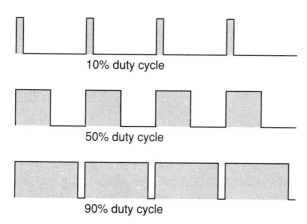

10% duty cycle

50% duty cycle

90% duty cycle

Figure 2-2:
Pulse Width Modulation

One of the problems with PWM is that instead of continuously supplying partial power, it will instead supply full power part of the time. In many cases this subtle difference has little overall effect. The LEGO MINDSTORMS motor, however, is an exception. These motors were designed to be very power efficient, and they have an internal flywheel that acts as a sort of energy storage tank. The flywheel is most effective when the motor has very little physical resistance (called load). Starting the motor consumes a lot of energy since the flywheel must also be spun up to speed. Once running, however, a lightly loaded motor (perhaps a motor that is turning a gear that isn't connected to anything else) can maintain its speed with very little external power. In this case, PWM does not affect the motor very much since the short pulses keep the flywheel spinning, and the flywheel itself keeps the motor spinning when the pulses stop. Under a heavy load (propelling a vehicle across a carpeted surface), the flywheel is less effective, and changing power levels will have a more noticeable effect on speed.

SENSORS

The RCX has three sensor ports, each of which can accommodate one of four different LEGO sensors: touch sensor, light sensor, rotation sensor, and temperature sensor. Sensors are connected to the RCX using the same type of wiring used to connect motors to the RCX output ports.

Each type of sensor has unique requirements for reading and interpreting its values; therefore the RCX must configure each sensor port before it can be used. There are actually two different settings configured: the *sensor type* and the *sensor mode.* The sensor type determines how the RCX interacts with the sensor; for example, the RCX will passively read a touch sensor, but must supply power to a light sensor. In general, the sensor type should match the actual type of sensor attached (touch, light, rotation, or temperature).

The sensor mode tells the RCX how to interpret a sensor's values. Some programming languages, such as RCX Code, automatically set the sensor mode based on the sensor type. Other languages, such as NQC, provide the additional flexibility of using any sensor mode with any sensor type (although some combinations are of little practical value).

Every sensor has three separate values associated with it: the *raw value, boolean value,* and *processed value.* The raw value is the actual reading of the sensor's voltage level, converted to a digital number from 0 to 1023 (inclusive). Every 3ms, the RCX reads the raw value for the sensor, then converts it to both a boolean value and a processed value.

Boolean Values

A sensor's boolean value can have one of two values: 0 or 1. Boolean values have their most obvious uses with a touch sensor, but occasionally other sensors can make use of boolean values as well.

The sensor mode includes a special parameter, called the *sensor slope*, which can range from 0 to 31 and determines how the raw value is converted into a boolean value. When the slope is zero, the RCX uses the conversion shown in Table 2–1.

Table 2-1:
Default Boolean Conversion

Condition	Raw Value
raw > 562	0
raw < 460	1
$460 \leq raw \leq 562$	unchanged

Note that a high raw value results in boolean value of 0, while a low raw value is a boolean value of 1.

The cutoff points (460 and 562) represent approximately 45% and 55% of the sensor's maximum value. It is not unusual for the raw values to bounce around by a few points; thus if a single cutoff was used, the boolean value would be susceptible to lots of jitter if the raw value was hovering near the cutoff. This method of reducing the amount of jitter in a boolean signal is called *hysteresis*.

When the slope parameter is non-zero, a different boolean conversion is used. Each time the sensor is read, its raw value is compared to the previous raw value. If the absolute value of this difference is less than the slope parameter, then the boolean value remains unchanged. If this difference exceeds the slope parameter, then the boolean value will be set to indicate whether the raw value increased (boolean value of 0) or decreased (boolean value of 1). Special cases exist at the extremes of the raw value range. This conversion is summarized in Table 2–2.

Condition	Boolean Value
change > slope	0
change < -slope	1
current > (1023-slope)	0
current < slope	1

Table 2-2:
Boolean Conversion
with Slope Parameter

slope = value of the slope parameter
current = current raw value
change = current value minus previous value

For example, consider the case where the slope is 10, and the initial raw value is 1020. This will result in a boolean value of 0. Let's say the raw value slowly decreases until it is 300. Even though a value of 300 is below the 45% threshold normally used for boolean values, the slope parameter causes the cutoff to be ignored and the boolean value will remain 0. If the raw value suddenly decreases, perhaps to 280 in a single reading, then the boolean value will become 1.

When set properly, the slope parameter can be used to configure a light sensor to ignore moderate variations and detect only abrupt changes. For example, the following NQC program configures sensor 2 to be a light sensor in boolean mode with a slope of 10 (more information on sensor types and modes appears later in the chapter). The program plays a high-pitched tone whenever the sensor rapidly goes from dark to light (sensor value equals 1) and a low-pitched tone during a rapid transition from light to dark (sensor value equals 0).

```
task main()
{
    SetSensorType (SENSOR_2, SENSOR_TYPE_LIGHT);
    SetSensorMode (SENSOR_2, SENSOR_MODE_BOOL + 10);

    while(true)
    {
        until (SENSOR_2 == 1);
        PlayTone (880, 10);
        until (SENSOR_2 == 0);
        PlayTone (440, 10);
    }
}
```

If you are using NQC and feel like experimenting with the slope parameter, then download this program to the RCX, attach a light sensor to sensor port 2, and run the program. Aim the sensor directly into a bright light, then try blocking and unblocking the light by covering the sensor with your finger. Rapid changes from light to dark (or vice versa) will cause the sensor's boolean value to change and the program will play a tone. Gradual changes such as slowly turning the sensor away from the light will not result in a change to the sensor's boolean value.

Small triangles on the RCX's display (below the sensor ports) indicate when a sensor's boolean value is 1.

Sensor Modes

There are 8 different sensor modes, some of which only make sense for certain sensor types. For each mode, a brief description of how it converts raw values into processed values is given below. The default modes (those automatically selected within RCX Code programs) for each sensor type are also indicated.

Raw Mode

Raw mode is the simplest mode of all. In this mode the processed value of a sensor is always equal to its raw value (an integer between 0 and 1023).

Boolean Mode

When boolean mode is selected, the sensor's processed value is set to the boolean value determined by the conversion described previously. For sensors other than the touch sensor, a proper slope parameter is often essential for meaningful boolean values. This is the default mode for a touch sensor.

Edge Count Mode

When edge count mode is selected, the RCX counts how many times the boolean value changes value. This count is initially 0 and increments by 1 each time the boolean value changes from 0 to 1 or from 1 to 0. This mode is only useful when the boolean value is meaningful.

Mechanical switches (such as the touch sensor) tend to chatter a bit; that is they will rapidly turn on and off several times in the process of being pressed or released. The RCX uses a *de-bounding* process to filter out the chatter. This de-bounding causes the RCX

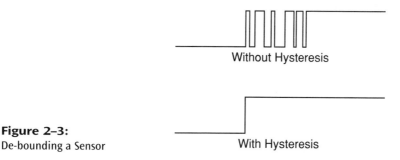

Without Hysteresis

With Hysteresis

Figure 2–3:
De-bounding a Sensor

to ignore edges for 300ms after a transition occurs. This increases the reliability of edge counting for most purposes, but also limits it to detecting edges that are at least 300ms apart.

Pulse Count Mode

The pulse count mode is similar to the edge count mode, except that the counter is only incremented when a boolean value goes from 1 to 0. Like edge counting, the boolean value is de-bounced to eliminate chatter. This mode is ideal for counting the number of times a button is pressed, for example.

Percentage Mode

In percentage mode, the raw value is converted into a value between 0 and 100 (higher raw values correspond to a lower percentage value). This is the default mode for a light sensor.

Rotation Mode

Rotation mode uses an algorithm that decodes the special output of the rotation sensor. This mode is the default for the rotation sensor, and it is meaningless when combined with any other sensor type. The resulting value from rotation mode is a cumulative rotation in increments of 22.5 degrees (a value of 16 represents a full revolution).

Celsius and Fahrenheit Modes

Celsius mode uses a special function to convert the raw value into a temperature reading. This function compensates for the specific characteristics of the temperature sensor. Fahrenheit mode uses the same function as Celsius mode to convert a raw value into a temperature, but then converts the temperature from Celsius to Fahrenheit.

Internally the RCX multiplies temperature values by 10, so that a value of 22.5 degrees is represented as 225. The LCD, however, will still display the actual temperature (e.g., 22.5). A preference setting in RCX Code development software determines whether Celsius or Fahrenheit mode is used as the default for temperature sensors.

Sensor Types

Touch Sensor

The touch sensor is the simplest of the LEGO sensors; it consists of a small pushbutton built into the end of a 2x3 brick. Multiple touch sensors may be connected in parallel—this is when more than one touch sensor is connected to the same input sensor port on the

Figure 2–4:
Touch Sensor

RCX. When used this way, the RCX would be able to detect when any of the sensors were pressed, but would not be able to distinguish which of the sensors was actually pressed. The touch sensor is often used in boolean mode, although edge counting and pulse counting can also be useful.

Care needs to be taken when attaching a wire to a touch sensor. Only the front four "nubs" have electrical contacts in them. If the wire is over these front contacts, then it can be attached in any direction. However, if the wire connector only overlaps two of the contacts, then it must be in the lengthwise direction as shown below:

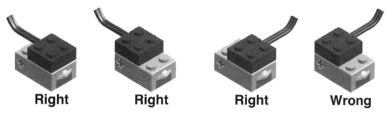

Right   **Right**   **Right**   **Wrong**

Figure 2-5:
Wiring a Touch Sensor

Light Sensor

The light sensor consists of a red Light Emitting Diode (LED) and a phototransistor that responds to incoming light. It is an active sen-

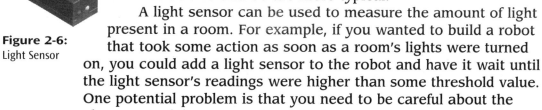

Figure 2-6:
Light Sensor

sor, which means it requires power from the RCX in order to operate properly. The light sensor is typically used in percentage mode, so its values may range from 0 to 100, although values between 40 and 60 are more typical.

A light sensor can be used to measure the amount of light present in a room. For example, if you wanted to build a robot that took some action as soon as a room's lights were turned on, you could add a light sensor to the robot and have it wait until the light sensor's readings were higher than some threshold value. One potential problem is that you need to be careful about the placement of the sensor since shadows or a small nearby light could fool the light sensor into believing the room is either darker or lighter than it really is.

Another potential problem is that light sensor readings can vary significantly between two different sensors, or even from the same sensor under different conditions. The amount of background light, colors and textures of nearby surfaces, and even the condition of the RCX's batteries can all affect the sensor's reading. Because of this, most programs that use a light sensor will need some adjustment in order to run properly. As the batteries in the RCX get weaker (or old batteries are replaced with new ones), programs that use the light sensor are likely to need further adjustment.

The light sensor responds to light coming from a fairly wide angle. Sometimes it is desirable to limit its detection to light coming from a specific direction. An example would be a robot that wanted to look around and find a bright light. This can be done quite easily

Figure 2-7:
Narrowing the Light Sensor's Field of Vision

by using one or more 1x2 beams to narrow the sensor's field of vision. Black beams work best since they reflect the least amount of light. It is important that the sensor's phototransistor line up with the beam's hole, which is easily accomplished by making the beam and the sensor's bottom flush with one another.

The previous examples used the light sensor to detect some external light source. Another use for the light sensor is to determine the color of some nearby surface. For example, a robot that sorts bricks by color needs to determine the color of each brick. Dependence on the ambient light in a room can make such a decision very difficult. Fortunately the light sensor contains its own light source (the LED), which allows it to measure reflected light from a nearby surface rather than taking a chance on the room's lighting conditions. For best results, the light sensor should be placed as close as possible to the target surface since this minimizes the influence of external light.

With a proper slope parameter, the boolean-based modes (boolean, pulse count, and edge count) may also be used with a light sensor to detect sudden increases or decreases in light.

Rotation Sensor

The rotation sensor is an optional sensor that is not included in the Robotics Invention System. It is an active sensor that only has a meaningful value when used in rotation mode.

Unlike the other sensors, which measure things on an absolute scale, the rotation sensor measures relative rotation (its readings are always relative to some base position). By default, the position when the program is started is used as the base position (rotation of 0), but a special command

Figure 2-8:
Rotation Sensor

allows you to reset the sensor so that the current position gets used as the base position.

In the base position, the sensor's value is 0. Turning the sensor in one direction will increase the sensor value, while turning it in the other direction will decrease the value. The sensor measures in increments of 22.5 degrees; thus a full revolution in one direction is equivalent to a sensor value of 16.

Gear reduction can be used to increase the accuracy of measurement. For example, a worm gear meshed with a 24-tooth gear provides 24:1 gear reduction. Placing the angle sensor on the worm gear axle yields 16x24 = 384 increments per revolution of the 24-tooth gear. The resulting accuracy is better than 1 degree.

Figure 2-9:
Increasing the Accuracy of a Rotation Sensor

Due to technical reasons, the RCX must closely monitor the rotation sensor, otherwise it may miss some rotation and yield incorrect readings. The RCX will reliably measure rotation at speeds up to 1250 rpm (revolutions per minute). This is more than fast enough for the standard LEGO MINDSTORMS Motor, which spins at about 350 rpm when unloaded.

Temperature Sensor

The temperature sensor is another optional sensor that is not part of the Robotics Invention System. Temperature readings can be made in either Celsius or Fahrenheit degrees. The operational range for the sensor is -20° C to +70° C (-3° F to +157° F). Temperatures outside this range are displayed as 999.9.

Figure 2-10:
Temperature Sensor

Due to the slow response of the temperature sensor, it is unsuitable for boolean use with a slope parameter. The temperature sensor also exhibits what is called a *nonlinear* response. What this means is that if you plotted the raw sensor reading versus temperature it would not appear as a straight line. The Celsius and Fahrenheit modes know how to correct for the nonlinearity, but raw and percentage modes do not.

As a result, the raw and percentage modes are not very useful with a temperature sensor.

Because the rotation and temperature sensors are not included in the Robotics Invention System, their use is minimized within this book. The rotation sensor is only used once in chapter 10. The temperature sensor is used for part of chapter 18. For more information about obtaining these sensors consult Appendix B.

OTHER CAPABILITIES

The RCX also provides an assortment of other capabilities including timers, variables, and a datalog. A brief description of these additional capabilities follows.

Keeping Time

The RCX provides a system clock, called the watch, which keeps track of the number of minutes since the RCX was turned on. In the default display mode, the LCD shows the current watch time in hours and minutes. Although the watch's current time is available to a running program, most programs require more accurate timing information than can be provided by the watch.

The RCX also provides four *timers,* which measure time in increments of 100ms (1/10th second). The timers are running whenever the RCX is turned on. A program may reset each timer independently at any time. The timers themselves will wrap around to 0 after about 55 minutes.

Variables

The RCX provides 32 storage locations which can be used by a program to store global variables. Each location can store a single 16-bit signed integer (an integer in the range -32,768 to 32,767). Since the variables are global across all of a program's tasks, care must be taken to ensure that a program with multiple tasks uses the variables in a consistent manner. The locations are shared across all of the programs and are not reset when changing programs. This makes it possible for one program to store information that will later be used by another program.

Variables can be assigned nearly any kind of value including constants, timer values, sensor readings, or random numbers. The firmware also provides a set of basic math operations for variables: addition, subtraction, multiplication, division, bitwise AND, bitwise OR, sign, and absolute value.

DATALOG

The RCX provides a data logging capability. A program may create a datalog of a fixed size, then add data points to the log. The data points may be sensor readings, variables, timer values, or the system watch. The datalog may then be uploaded to a host computer for analysis or some other function.

Both the datalog and RCX programs must share the same 6K of user memory (6101 bytes to be exact). Since each data point requires 3 bytes of memory the largest possible datalog will hold 2033 points. However, a datalog this large would leave no room for any programs.

The RCX's display includes a datalog indicator which consists of a circle cut into four quadrants. If no datalog exists, then the indicator remains off. When a datalog is in use, one or more of the quadrants is shown, indicating roughly how much of the datalog is currently filled up. One of the quadrants will be flashing as long as there is still room for additional points. Once the datalog is full, all four quadrants will become solid.

AUTOMATIC POWER DOWN

In order to conserve batteries, the RCX will automatically turn itself off after several minutes of inactivity. The amount of time before the RCX turns itself off is called the *power down time* and is initially set to 15 minutes. It is worth noting that the RCX will turn itself off even while a program is running, so in order to run a program unattended for a long time it is necessary to change the power down time.

The specifics of changing this setting vary depending on the software used to program the RCX. To change this setting from within the MINDSTORMS Robotics Invention System software, start at the **Main Menu** screen, then select **Getting Started** and **Set Up Options.** Enter the desired number of minutes for power down time or select "infinity" to disable the automatic power down.

To change the power down time using the command line version of NQC, use the following command, substituting the desired number of minutes:

```
nqc -sleep minutes
```

Setting the power down time to 0 disables the automatic power down:

```
nqc -sleep 0
```

If you are using RcxCC, the power down time can be changed by selecting **Diagnostics** from the **Tools** menu and entering the appropriate time in the dialog box. As with NQC, a time of 0 minutes indicates that the RCX should never turn itself off.

CONCLUSION

This chapter discussed the different features of the RCX including tasks, motor outputs, sensors, and timers. The next chapter discusses how to use these features from within NQC. Then, starting with chapter 5, a series of robots are presented, each demonstrating how these features can be combined to make robots that carry out predefined tasks and respond to their environment.

Chapter 3
Introduction to NQC

This chapter provides an introduction to the NQC programming language. Many of the more commonly used NQC features are presented. For a complete treatment of NQC syntax and commands refer to the *NQC Programmer's Guide*.

It is assumed that the reader has already installed NQC (or an NQC-based development environment such as RcxCC) and has already gone through the basic process of creating, compiling, and downloading a program as described in chapter 1.

A SIMPLE PROGRAM

Our first program assumes that there is a motor attached to output A and a touch sensor on input 1. When run, the program will turn on the motor, wait until the touch sensor is pressed, then stop the motor.

```
// intro_1.nqc - a very simple program
task main()
{
    SetSensor(SENSOR_1, SENSOR_TOUCH);

    On(OUT_A);
    until(SENSOR_1 == 1);
    Off(OUT_A);
}
```

In general, the amount of white space (spaces, tabs, and carriage returns) doesn't matter in a program, but punctuation is critical. The compiler is also case sensitive, so "On" is not the same as "on."

The first line of the program is a comment. Comments are ignored by the compiler and allow the programmer to provide explanations or annotations to the code itself. There are two types of comments. A single-line comment starts with two forward slashes (//) and causes the compiler to ignore the remainder of the line.

```
// this is a single-line comment

/ NOT a comment - only one slash
// NOT a comment - space between the slashes
```

The second type of comment can span multiple lines. It starts with /* and ends with */.

```
/* comment */ no longer a comment

/* this is a multiline comment...
    still a comment...
    last line */

/* start the comment
    /* <--this is ignored--no nesting
    end the comment --> */
    we are no longer in the comment
*/ <--NOT part of the comment--already ended
```

Which type of comment you use is largely a matter of personal style. I prefer to use the // form most of the time. This allows the use of /* */ to comment out large pieces of code (with their embedded // comments) during testing and debugging.

Programs are built from code blocks (*tasks*, and *subroutines*), each of which starts with its type (task, or sub), followed by its name and a pair of parentheses ("(" and ")"). The code itself is enclosed in braces ("{" and "}"). The intro_1.nqc program shown above contains a single task, named "main," whose structure looks like this:

```
task main()
{
//...body of the task...
}
```

Every program must contain a task named "main"; this is the task that is activated when the program is run. The other tasks may be named however the programmer desires. Names in NQC must start with a letter or the underscore ("_"). Remaining characters in the name can be a letter, a numeric digit, or an underscore. For tasks and subroutines, the name must be unique; that is, you cannot use the same name to refer to two different code blocks. NQC also reserves a number of names (called keywords) for its own use. A complete list of keywords appears in appendix D.

A code block (such as a task) contains a list of statements. Each statement is an instruction for the program to do something. Statements must always end with a semicolon (";").

The RCX must be told what sort of sensor is attached to each sensor port. This is done by the SetSensor command, which takes two arguments: the sensor port and the sensor configuration. NQC uses the names SENSOR_1, SENSOR_2, and SENSOR_3 for the RCX's sensor ports. Other names (such as SENSOR_TOUCH) are used to define the different sensor configurations. In this case, sensor 1 is configured for a touch sensor.

```
SetSensor(SENSOR_1, SENSOR_TOUCH);
```

The next statement uses the On command to turn on a motor. NQC uses the names OUT_A, OUT_B, and OUT_C for the RCX's output ports. In this case, output A is turned on.

```
On(OUT_A);
```

Our previous statements were both examples of commands, which always allow execution to continue on to the next statement once their work is complete. Control structures, on the other hand, allow constructs such as loops and conditional tests to be written. Our next statement uses the until control structure, which waits until a condition is true before proceeding. The condition itself must be enclosed in parentheses—"(" and ")." In this case our condition is that the value of sensor 1 (which is the touch sensor) must be equal to 1 (which means it is pressed). When the RCX gets to this statement it will keep checking sensor 1 until the condition is met, and only then will it move on to the next statement.

```
until(SENSOR_1 == 1);
```

Our last statement uses the Off command to turn the motor off.

```
Off(OUT_A);
```

Although our program was very simple, it was built from the same basic elements that all NQC programs are constructed from: tasks, statements, control structures, and conditions. The following sections will explain all of these capabilities in more detail as well as present some of the more advanced features of NQC.

CONTROLLING OUTPUTS

In the previous section we used the On and Off commands to start and stop a motor attached to output A. Since controlling motors (or other output devices such as lights) is one of the primary activities of a program, NQC provides a variety of commands for this.

The commands to control an output's mode are On(*outputs*), Off(*outputs*), and Float(*outputs*). Each of these commands takes a single argument, which must be a set of outputs. NQC uses the names OUT_A, OUT_B, and OUT_C to refer to the three outputs. Multiple outputs may be combined with the + operator, so that a single command can control multiple outputs simultaneously.

```
On(OUT_A + OUT_B);    // turn on outputs A and B
Off(OUT_A);           // turn off A
Float(OUT_B);         // let B "coast"
```

Turning an output on is self-explanatory, but turning an output off is a little more complicated. Take the example of an output that is used to power a motor that in turn propels a vehicle forward. Once the vehicle has started moving it has some momentum. Even if power is no longer applied to the motor, the vehicle may continue to coast for a bit. This is similar to what happens when one is riding a bicycle and stops pedaling. In NQC, this is accomplished by making an output float. It is also possible to control a motor so that it resists any further motion. Returning to the bicycle analogy, this is like applying the brakes. In NQC, this is done by turning an output off.

When an output is on, it can be running in one of two directions: forward or reverse. There are three commands to control direction: Fwd(*outputs*), Rev(*outputs*), and Toggle(*outputs*), each of which takes a set of outputs as an argument. Fwd and Rev set the direction to forward and reverse, respectively. Toggle changes the direction to the opposite of whatever it was previously. Note that these commands can be used even when the output is off (or floating); the direction will then be remembered the next time the motor is turned on.

```
Fwd(OUT_A);              // set A to the forward direction
Rev(OUT_B);              // set B to the reverse direction
Toggle(OUT_A + OUT_B);   // toggle A and B
```

Turning a motor on and setting its direction are so often done together that NQC provides commands combining these actions.

```
OnFwd(OUT_A);            // turn on A in forward direction
OnRev(OUT_B);            // turn on B in the reverse direction
```

These commands are merely for the programmer's convenience. The above two statements are functionally identical to the following sequence of statements:

```
// OnFwd(OUT_A)...
Fwd(OUT_A);
On(OUT_A);
// OnRev(OUT_B)...
Rev(OUT_B);
On(OUT_B);
```

It is also common to turn on a motor for a specified amount of time, then turn it off. The OnFor(*outputs, duration*) command is a convenient way to do this. The duration is specified in hundredths of a second.

```
// turn on OUT_A for 1 second
OnFor(OUT_A, 100);

// same thing, but without using OnFor
On(OUT_A);
Wait(100);
Off(OUT_A);
```

The RCX is also capable of changing the power level of each output. Like direction, the power setting can be set when an output is off and will be remembered for the next time the output is turned on. The command for setting an output's power level is SetPower(outputs, power). Outputs are specified just like in all of the other output commands. The power level should be an integer between 0 and 7, where 0 is the lowest power level and 7 is the highest. NQC also defines the names OUT_LOW and OUT_FULL to represent minimal and full power respectively.

```
SetPower(OUT_A + OUT_B, 0);   // set lowest level for A and B
SetPower(OUT_C, 7);           // set highest level for C
SetPower(OUT_C, OUT_FULL);    // set highest level for C
```

> When displaying a power level, the RCX adds one to the actual power level; thus the lowest power level is displayed as 1 and the highest as 8.

MISCELLANEOUS COMMANDS

The Wait(time) command causes the task to pause for a specified amount of time, which is measured in hundredths of a second.

```
Wait(200);     // pause for 2 seconds
```

The PlayTone(frequency, duration) command plays a single tone of specified frequency and duration. Both the frequency and duration must be constants. The duration is measured in hundredths of a second.

```
PlayTone(440, 25);    // play a note for 1/4 second
```

The PlaySound(*number*) command plays one of the RCX's six built-in sounds. The number argument must be a constant between 0 and 5 inclusive. NQC also defines constants representing each of the sounds.

```
PlaySound(SOUND_CLICK);      // play a sound
```

Name	Number	Type of Sound
SOUND_CLICK	0	short click
SOUND_DOUBLE_BEEP	1	two beeps
SOUND_DOWN	2	descending sweep
SOUND_UP	3	ascending sweep
SOUND_LOW_BEEP	4	very low tone
SOUND_FAST_UP	5	fast ascending sweep

Table 3-1:
Sounds

The ClearTimer(*number*) command resets a timer's value to 0. The number argument must be a constant between 0 and 3 inclusive, and it specifies which of the four timers to reset. The timers themselves may be read using the Timer(*number*) expression (see the Expressions section for more information).

```
ClearTimer(0); // reset timer 0
```

USING #DEFINE

In our first example we referred to the sensor and motor by the name of the sensor or output port they were attached to (SENSOR_1 and OUT_A). For such a simple program this was fine, but for larger programs it can be a little cumbersome to always remember which sensor and which output to use. Wouldn't it be nice if we could assign our own names to the input and output ports?

NQC provides a preprocessor which allows you to do just that. As its name suggests, the preprocessor works on your program before handing it to the compiler proper. Commands for the preprocessor are called directives and always start with a pound sign ("#"). One of the most powerful directives is #define, which allows you to assign a name to an arbitrary sequence of tokens. Such a definition is called a *macro*. For example, the following line creates a macro called CRANE_ARM that is defined as OUT_A.

```
#define CRANE_ARM     OUT_A
```

Whenever it sees the macro's name, the preprocessor will act as if the definition were used instead. For example, if the program contains

```
On(CRANE_ARM);
```

the preprocessor will act as if the program contained

```
On(OUT_A);
```

Using #define to create more meaningful names for the sensors and outputs makes programs easier to read. It also makes them easier to modify. Let's say you were changing the way your robot was built and in the process moved the crane arm motor from output A to output B. Without a #define, you may have had many references to OUT_A sprinkled throughout your program—all of which would have to be changed to OUT_B. If a #define was used as shown above, and the rest of the program used CRANE_ARM, rather than referring to OUT_A directly, then the program could be adapted to the new robot simply by changing the macro:

```
#define CRANE_ARM     OUT_B
```

Although the previous example simply defined one name as a synonym for another, macros work just as well if their definition contains several items. For example:

```
#define ALL_MOTORS    OUT_A + OUT_B + OUT_C
On(ALL_MOTORS); // turns on all motors!
```

Macros can also be defined with a list of arguments. The argument list appears within parentheses, and a comma (",") is used to separate the argument names. For example, the following macro turns one or more outputs on for a specified number of seconds:

```
#define OnForSeconds(outs, secs)   OnFor(outs, (secs)*100)
```

OnForSeconds is a fairly simple macro—all it needs to do is multiply the number of seconds by 100 and then call OnFor.

When defined with an argument list, the macro must be accompanied by the correct number of arguments wherever it is used. This applies even if the #define has an empty argument list.

```
#define OnForSeconds(outs, secs) OnFor(outs, (secs)*100)

OnForSeconds(OUT_A, 5);     // ok, turns on A for 5 seconds
OnForSeconds;               // ERROR—no argument list!
OnForSeconds(OUT_A);        // ERROR—not enough arguments!

#define OnA()               On(OUT_A)

OnA();                      // ok, turns on A
OnA;                        // ERROR—no argument list!
```

The definition of OnForSeconds contained parentheses around the use of secs. Without these parentheses the macro could result in some surprising behavior. Consider the following:

```
#define OnForSeconds(outs, secs)   OnFor(outs,  secs*100)

OnForSeconds(OUT_A,  2+3);
```

At first glance it would appear that output A would be turned on for 5 seconds; however, using a macro results in literal substitution of its arguments into the definition. In this case, "OUT_A" would be substituted for "outs" and "2+3" would be substituted for "secs," resulting in the following:

```
OnFor(OUT_A,  2+3*100)
```

Because multiplication has higher precedence than addition, the above call will turn on the output for 3.02 seconds rather than 5 seconds. Using parentheses within the macro definition avoids this potential problem. An even better solution is to use functions (described later in the chapter) instead of macros whenever possible.

Even so, the macro facilities are so useful that many of the commands and names that NQC provides are actually macros defined internally within the NQC compiler. For example, the until control structure is implemented using the following macro:

```
#define until(cond)   while(!(cond))
```

USING SENSORS

The RCX has three sensor ports (labeled 1, 2, and 3), each of which can be connected to a sensor. The RCX has no way of knowing what type of sensor is connected to each sensor port, so an NQC program must contain statements to configure the sensor ports. The primary command for this is SetSensor(*sensor_port*, *sensor_config*). In NQC, sensor ports are named SENSOR_1, SENSOR_2, and SENSOR_3 (more meaningful names can be assigned using macros as described above). The value used for *sensor_config* consists of both sensor type and sensor mode information. NQC predefines the following type/mode combinations:

Configuration	Type	Mode
SENSOR_TOUCH	Touch	Boolean
SENSOR_LIGHT	Light	Percentage
SENSOR_ROTATION	Rotation	Rotation
SENSOR_CELSIUS	Temperature	Celsius
SENSOR_FAHRENHEIT	Temperature	Fahrenheit
SENSOR_PULSE	Touch	Pulse Count
SENSOR_EDGE	Touch	Edge Count

Table 3-2:
Sensor Configurations

For example:

```
// sensor 1 is a touch sensor
SetSensor(SENSOR_1,  SENSOR_TOUCH);
// sensor 2 is a light sensor
SetSensor(SENSOR_2,  SENSOR_LIGHT);
```

Additional information on the sensor types and modes themselves was presented in chapter 2.

Several sensor modes (Rotation, Edge Count, and Pulse Count) measure relative quantities and can be reset using the ClearSensor(*sensor_port*) command.

```
SetSensor(SENSOR_3,  SENSOR_ROTATION);
ClearSensor(SENSOR_3);       // reset rotation to 0
```

Unlike the outputs, multiple sensor ports cannot be combined into a single command.

```
SetSensor(SENSOR_1 + SENSOR_2,  IN_TOUCH);      // ERROR!
```

Sensor type and mode may be set individually using the SetSensorType(*sensor_port, sensor_type*) and SetSensorMode(*sensor_port, sensor_mode*) commands. The sensor slope (if desired) may be added to the sensor_mode parameter.

```
// setup sensor 2 as a boolean light sensor with slope 10
SetSensorType(SENSOR_2, SENSOR_TYPE_LIGHT);
SetSensorMode(SENSOR_2, SENSOR_MODE_BOOLEAN + 10);
```

A sensor is read simply by using its name within an expression. For example, the simple program at the start of this chapter used the following statement to wait for a touch sensor to be pressed:

```
until(SENSOR_1 == 1);
```

Further details on expressions and conditions are given later in the chapter.

USING VARIABLES

NQC uses the RCX's 32 storage locations as variables, which may be declared either globally or locally (within a task, function, or other code block). Globally declared variables are accessible to the entire program, whereas locally scoped variables are only available within the code block that defined them. In both cases the variable must be declared prior to its first use. A variable declaration is the keyword int, followed by a comma-separated list of variable names, and terminated by a semicolon (";"). Optionally, a variable may be assigned an initial value in the declaration.

```
int foo;           // global variable foo

task main()
{
    int bar, baz; // bar and baz are local to main
    int initMe = 1;   // initMe is initialized to 1
}
```

In some cases, such as within complicated expressions or conditions, NQC must allocate one or more temporary variables for its own use. In general, NQC tries to re-use these temporary variables as much as possible, but they still deplete the total number of locations available for your own variables. If the number of required storage locations (for variables and temporary storage) exceeds the 32 locations provided by the RCX, NQC will report an error.

Variables may be assigned a value using an assignment statement.

```
int bar;

bar = 7; // set the value of bar to 7
```

There are also several variants of the assignment operator that allow a variable to be operated on arithmetically.

```
bar += 7;       // add 7 to bar
bar -= 3;       // subtract 3 from bar
bar *= 5;       // multiply bar by 5
bar /= 2;       // divide bar by 2
```

Although the above examples used constant integers on the right-hand side of the assignment, any legal expression may be used (see below).

EXPRESSIONS

The simplest expression is an integer constant. Constants are interpreted as decimal values unless they are preceded by "0x" in which case they are interpreted as hexadecimal values.

```
bar = 10;        // set the value of bar to 10
bar = 0xa;       // also set the value of bar to 10
```

Variable names and sensor port names are also legal expressions, and they refer to the value of the variable or sensor. Additionally, the expression Random(x) uses the RCX's random number generator to pick a number between 0 and x (inclusive).

```
bar = baz;       // assign the value of baz to bar
bar = SENSOR_1;  // assign the value of sensor 1 to bar
bar = Random(4); // assign a number between 0 and 4 to bar
```

Two other expressions provide access to the RCX's timers and message capability.

```
bar = Timer(0);  // assign value of timer 0 to bar
bar = Message(); // read the last received IR message
```

Expressions may be combined using several different arithmetic operations.

```
bar = 1 + Random(4);  // random between 1 and 5
bar = SENSOR_1 - 7;   // sensor 1 minus 7
bar = SENSOR_1 * 4;   // sensor 1 times 4
bar = SENSOR_1 / 3;   // sensor 1 divided by 3
bar = baz & 0x7;      // bitwise AND
bar = baz | 0x4;      // bitwise OR
```

There are also three unary operators.

```
bar = - baz;      // negation
bar = abs(baz);   // absolute value
bar = sign(baz);  // sign (-1, 0 or +1)
```

Constant expressions may use several additional operators. It is an error to apply these operators to nonconstant expressions such as variables or input sensors.

```
bar = 100 % 7;    // 100 modulo 7
bar = 7 << 2;     // left shift 7 by 2 bits
bar = 100 >> 1;   // right shift 100 by 1 bit
bar = 5 ^ 4;      // bitwise exclusive OR
bar = ~0x80;      // one's complement (bitwise negation)
```

Two special operators, ++ and --, provide the ability to increment and decrement a variable. Since they implicitly modify their operand, they can only be used on variables. In addition, the operators may be applied as prefix or postfix operations, meaning they can take effect before or after the value of the variable is used.

```
baz = 1;
bar = baz++;      // postfix: bar=1, baz=2
bar = ++baz;      // prefix: bar=3, baz=3
baz--;            // baz=2
bar = 123++;      // ERROR--123 cannot be incremented!
```

The precedence and associativity of all of the operators follow that of C, and this is summarized in Appendix D. Parentheses may be used to change the order of computation.

```
bar = 2 + 3 * 4; // bar = 14
bar = (2 + 3) * 4;   // bar = 20
```

CONDITIONS

Our sample program needed to wait until the touch sensor was pressed. This was done with the until statement, which waits until a specified condition becomes true.

```
until(SENSOR_1 == 1);
```

Most conditions involve comparing one expression with another; in the above example the value of sensor 1 was compared to the constant 1. There are six different relational operators that can be used in a condition: ==, !=, <, >, <=, and >=. An example of each of these, along with a description of the condition to be met, is shown below.

```
until(SENSOR_1 == 1);    // sensor 1 equal to 1
until(SENSOR_1 != 1);    // sensor 1 not equal to 1
until(SENSOR_1 < 10);    // sensor 1 less than 10
until(SENSOR_1 > 20);    // sensor 1 is greater than 20
until(SENSOR_1 <= 15);   // sensor 1 less or equal to 15
until(SENSOR_1 >= 25);   // sensor 1 greater or equal to 25
```

Of course, the expressions in a condition aren't limited to sensors and constants.

```
until(SENSOR_1 > bar);   // use the value of bar
until(Timer(0) < 100);   // wait for timer
```

Conditions may be combined using && (logical AND) or || (logical OR). Note that these operators are different from the bitwise AND and bitwise OR operators for expressions (&, and |). Bitwise operators act on the individual bits in a variable and are not suitable for use in combining conditions.

```
// wait until SENSOR_1 between 10 and 20
until( SENSOR_1 > 10 && SENSOR_1  <  20);

// wait until either SENSOR_1 or SENSOR_2 equals 1
until (SENSOR_1 == 1 || SENSOR_2 == 1);
```

Conditions may also be negated with the ! (logical negation) operator. This operator has a higher precedence than && and ||, but parentheses may be used to change the grouping.

```
until (! SENSOR_1 == 1);  // wait until sensor 1 does NOT equal 1

// wait until neither SENSOR_1 or SENSOR_2 are equal to 1
until (! (SENSOR_1 == 1 || SENSOR_2 == 1));
```

Finally, there are two constant conditions: true and false.

```
until(false);   // wait forever
until(true);    // no wait at all...always true
```

CONTROL STRUCTURES

Control structures can be used to make decisions or create conditional and unconditional loops. Most control structures make use of a condition, which must always be enclosed in a set of parentheses. The if statement allows a statement to be executed if a condition is true.

```
// turn off output A if sensor 1 equals 1
if (SENSOR_1 == 1)
    Off(OUT_A);
```

The indenting makes the statement easier to read but has no effect on its execution. Only the statement immediately following the condition is considered part of the if structure. If multiple statements need to be executed conditionally, they should be grouped together using braces ("{" and "}"). This concept of grouping multiple statements together with braces that are to be treated as a single statement applies to all of the control structures.

```
// not quite right...
if (SENSOR_1 == 1)
    Off(OUT_A); // this is executed conditionally
    On(OUT_B); // this is executed unconditionally

// better--both Off() and On() are conditional
if (SENSOR_1 == 1)
{
    Off(OUT_A);
    On(OUT_B);
}
```

An if condition may also have an else clause, which is executed if the condition is false.

```
if (SENSOR_1 == 1)
    On(OUT_A);  // executed if SENSOR_1 equals 1
else
    On(OUT_B);  // executed if SENSOR_1 does not equal 1
```

The while statement causes a statement (or statements) to be repeatedly executed so long as the condition is true. If the condition is initially false, then the body of the while statement will not be executed at all.

```
// keep reversing output A as long as sensor 1 equals 1
while(SENSOR_1 == 1)
    Toggle(OUT_A);
```

The do while statement is the same as the while statement, with the exception that the condition is checked after the body instead of before. As a result, the body will always be executed at least once.

```
do
{
    Toggle(OUT_A);
} while(SENSOR_1 == 1);
```

The repeat statement executes a statement a specified number of times. Unlike the previous control structures, it uses an expression rather than a condition. The expression is evaluated a single time, and the result of that expression is used to determine the number of repeats.

```
// repeat 10 times...
repeat(10)
{
    Toggle(OUT_A);
}
```

USING TASKS

The main task must always exist—it is the task that is started when a program is run. An NQC program may contain up to nine additional tasks, which are defined by the same syntax as the main task, but have unique names.

```
task main()
{
    // this is the main task
}

task foo()
{
    // this is another task
}
```

When a program is run, only the main task is active. The other tasks may be started or stopped using the start and stop statements. The example below starts two tasks, then stops one of them 10 seconds later.

```
// intro_2.nqc - using multiple tasks

task main()
{
    start fast_beep;
    start slow_beep;
    Wait(1000);
    stop slow_beep;
}

task fast_beep()
{
    while(true)
    {
        PlayTone(440, 10);
        Wait(50);
    }
}

task slow_beep()
```

```
{
    while(true)
    {
        PlayTone(880, 10);
        Wait(100);
    }
}
```

Even though the main task ends (after stopping the slow_beep task), the program continues to run. This is because the fast_beep task is still active. In order to stop a program completely, all tasks must be stopped. This can be done with the StopAllTasks command.

```
StopAllTasks();
```

FUNCTIONS

So far, all of the example code has been very simple, and the statements for a task have appeared directly within the task's body. When writing larger programs, however, it is often desirable to break up a program into several smaller pieces. These pieces are called *functions* and make programs easier to read, maintain, and reuse. A function's definition looks similar to that of a task, but it uses the keyword void rather than task.

```
void onUntilPressed()
{
    On(OUT_A);
    until(SENSOR_1 == 1);
    Off(OUT_A);
}
```

The function can then be invoked from within a task (or another function).

```
task main()
{
    SetSensor(SENSOR_1,  SENSOR_TOUCH);
    onUntilPressed();
}
```

This example does the same thing as the intro_1.nqc example at the start of the chapter; specifically it turns on motor A until a touch sensor attached to input 1 is pressed. Functions really start to pay off when a piece of code needs to be executed more than once. For example, let's say we want to run motor A until the touch sensor is pressed, then we want to run motor B for one second, then resume A until the touch sensor is pressed again. We can use the previous onUntilPressed function and simply call it twice from the main task.

```
task main()
{
    SetSensor(SENSOR_1,  SENSOR_TOUCH);
    onUntilPressed();
    OnFor(OUT_B,  100);
    onUntilPressed();
}
```

Functions can also have one or more arguments, which are declared after the function name. Each argument is specified as a type followed by a name, and multiple arguments are separated by commas. For example, consider the following function, which turns output A on and off a specified number of times and for a specified duration:

```
void pulse(int count, int duration)
{
    repeat(count)
    {
        OnFor(OUT_A, duration);
        Wait(duration);
    }
}
```

The arguments are named count and duration and are both of type int. When calling a function, the appropriate number of arguments must be supplied. Using the above function, a program could send 5 pulses of 1 second each by making the following call:

```
pulse(5, 100);
```

Although all numbers within the RCX are the same data type (signed 16-bit integers), argument types are still required in order to determine how an argument is passed from the code calling the function (the caller) to the function itself (the callee).

The first type, int, is perhaps the most basic and was used in the preceding definition of pulse. It represents a temporary variable that gets used within the function. Argument passing is done by copying the value of the actual argument into this temporary variable. The function then can use this temporary variable however it wants. This is known as pass by value. There are two important features of passing by value: the actual argument cannot be altered by the function (it can only alter the temporary variable), and the value of the actual argument is latched into the temporary variable when the function is called. This second feature is of particular importance in NQC since arguments are often dynamic sources such as sensor values (e.g., SENSOR_1). The downside to passing by value is that it typically requires using up one of the 32 RCX storage locations to hold the temporary variable and can be less efficient than using one of the other types.

The second type, const int, carries with it the restriction that only constant expressions (numbers or arithmetic operations on constant expressions) may be used as actual arguments. There is no corresponding notion in C—this is purely an NQC invention due to some limitations in the RCX firmware. This argument type is especially useful when the argument needs to be used in a place where only constant values are allowed. For example, outputs are always identified by constant expressions, so a const int argument could be used to pass an output name to a function.

```
void pulse2(const int out, int count, int duration)
{
    repeat(count)
    {
        OnFor(out, duration);
        Wait(duration);
    }
}
```

This function could then be called like this:

```
pulse2(OUT_A,  5,  100);
```

If the out argument were of type int rather than const int, the function would not compile. This is because the OnFor command requires its first argument to be a constant expression, and if out were of type int the compiler would not be able to ensure that it is always constant. Many of the standard RCX commands (e.g., On, Off, Fwd) make extensive use of const int arguments.

The third type, int &, represents pass by reference. In this case, instead of allocating a temporary variable and copying the actual argument into it, the argument itself is used within the body of the function. This allows a function to modify the argument so that the caller will see the modified value. For example:

```
void swap1(int x, int y)
{
    int temp;
    temp = x;
    x = y;
    y = temp;
}

void swap2(int &x, int &y)
{
    int temp;
    temp = x;
    x = y;
    y = temp;
}
```

The definitions of swap1 and swap2 are identical, except for the types of the arguments. The intent behind both functions is to swap the values of the two arguments. The problem with swap1 is that it uses call by value, so when called like this

```
swap1(a, b);
```

the values of a and b will be copied into temporary variables. The contents of these temporary variables will then get swapped, but a and b will remain unchanged. By using pass by reference, swap2 avoids this problem. No temporary variables are created, and the function operates on its arguments directly. Note that it is illegal to use anything other than a variable as an actual argument for type int &.

```
swap2(a, 123); // ERROR, 123 is not a variable!
```

The fourth type, const int &, is a strange beast. It is passed by reference, but with the restriction that the item being passed will not be directly modified by the function. This means that the argument may not be used as the target of assignment or as the operand of an increment or decrement operation. In effect, this is a promise from the function to its caller that the value will not be changed. The most significant benefits to this are that any expression may be passed this way (not just variables), and it is often more efficient than an ordinary int argument. For example, the following function is like the original definition of pulse, with the exception that it uses const int & arguments.

```
void pulse3(const int &count, const int &duration)
{
    repeat(count)
    {
        OnFor(OUT_A, duration);
        Wait(duration);
    }
}
```

It can be called just like its predecessor, but unlike the original version no temporary variables are used to hold copies of the arguments.

```
pulse3(5, 100);
```

What argument type should be used? There are no hard and fast rules, but a couple of guidelines will suffice most of the time. If a constant expression is required (perhaps because the argument will in turn be used in a command like On), then the argument type must be const int. If the argument is to be modified by the function (and the modified value returned to the caller), then int & is required. In most other cases const int & is probably the right choice. The only reason to fall back on int is if a copy of the actual argument is desired.

The four different argument types are summarized in the table below.

Type	Semantics	Comments
int	by value	uses up a variable, actual argument is not altered
int &	by reference	efficient, actual argument must be a variable
const int	n/a	efficient, actual argument must be constant
const int &	by reference	efficient, argument cannot be modified

Table 3-3:
Argument Types

CONCLUSION

Although far from a complete reference on NQC, the material presented in this chapter serves well as an introduction to the commonly used NQC features. The robots in this book, starting with chapter 5, provide a more hands-on approach to learning NQC. While working with those sample programs it may be helpful to refer back to this chapter or the summary provided in appendix D.

These first few chapters have concentrated on the electrical and programming aspects of robots. Mechanical aspects are equally important, which is why chapter 4 introduces various techniques and principles used in actually building robots.

Chapter 4
Construction

This chapter describes the various mechanical principles and techniques used in constructing LEGO robots. A number of structures will be introduced, followed by examples of using special pieces such as gears, pulleys, and the differential. Builders who have spent years constructing LEGO Technic models will find many of the procedures to be second nature already. However, those unfamiliar with these special pieces will find many helpful hints in the material.

STRUCTURES

When building a robot it is tempting to think only of the motors, sensors, and gears that will make the robot come alive. The use of motors, however, subjects these creations to unusual forces, and without a solidly constructed body a robot may fall apart at the most inopportune moment. The following pages present several structures that can be used to make more robust robots.

Frames

When constructing a rectangular structure, it is often desirable to make it sturdy while still using a minimal number of pieces. Our first attempt, shown below, is about as simple as it gets. The frame itself is made up of 1x12 and 1x4 beams. To hold the beams together, 1x6 plates are used both above and below the 1x4 beams.

When discussing LEGO, it is convenient to use *LEGO units*, rather than actual measurements in inches or centimeters. After all, which is more descriptive: a 2x4 brick or a $^{10}\!/_{16}$" x 1¼" brick? A LEGO unit is equal to the distance between adjacent "studs" on an ordinary brick, which happens to be $^{5}\!/_{16}$".

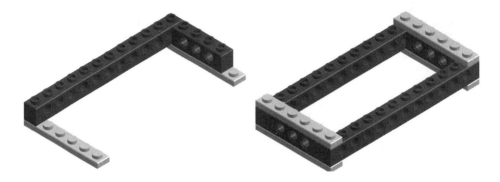

Figure 4–1:
Simple Frame

This structure stays together well enough, but it has a little too much wiggle to be of much practical use. The problem is that the 1x6 plates do not lock the corners at right angles. This can be remedied by using wider plates. In fact, a single 2x6 plate makes the entire structure much more rigid.

Figure 4–2:
Improved Frame

Long Beams

It is often necessary to build structures that are longer than the largest beam. The simplest way is to join one beam to another using plates (or bricks) above and below the seam. This arrangement is relatively compact and quick to build; however, it is not strong enough to support a large weight. Thus it would not be good for something like the arm of a crane.

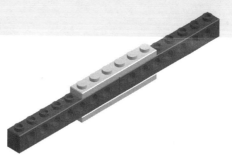

Figure 4–3:
Simple Long Beam

A more robust way to join two beam lengths is to use *friction pegs* to attach a third beam that overlaps the seam. The overlapped beam must be at least a 1x6 (in order to have room for all four friction pegs), and longer beams will result in a stronger structure. The placement of the friction pegs is important—they should be at the very ends of the beams in order to maximize stability.

Figure 4–4:
Better Long Beam

This arrangement is capable of supporting a significant weight, but it flexes slightly from side to side. If this is a concern, the structure can be reinforced by top and/or bottom plates.

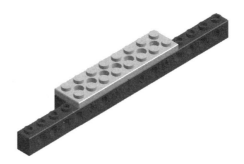

Figure 4–5:
Best Long Beam

Vertical Beams

So far, we have only used beams in a horizontal orientation. What if we need to attach them vertically? The friction pegs will allow us to connect beams at any angle, but when we try to put a second peg in the vertical beam we find that the holes don't quite line up.

Figure 4–6:
Vertical Miss

This is because of the unfortunate fact that a 1x1 brick is a little taller than it is wide. In fact, a brick is 1.2 LEGO units tall. Since three plates stack up to make a brick, a plate is 0.4 LEGO units tall. It just so happens that a stack of one brick and two plates is exactly 2 LEGO units tall (1.2 + 2x0.4 = 2.0). Taking this into account, we can build the following structure:

Figure 4–7:
Vertical Beam

The vertical beam arrangement is quite sturdy. In extreme cases, a third beam can be used to reinforce it further. The key is to form a right triangle with integral lengths for all three sides. The most common dimensions for such triangles are 3-4-5 and 5-12-13 (or multiples thereof).

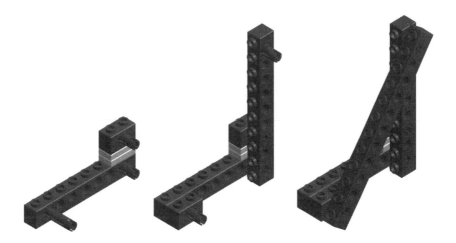

Figure 4–8:
Triangular Structure

Bracing

Vertical beams can also be used to provide bracing for a structure. The resulting arrangement is extremely strong. This can be used quite effectively on larger robots—vertical bracing from the top to the bottom of a robot ensures the integrity of everything in between because the brace will prevent any individual horizontal pieces from pulling apart.

Figure 4–9:
Vertical Bracing

Wide Beams

Let's say you need a small platform at the end of a long beam. In the example below, a small platform is created by using two 1x4 beams alongside the main 1x12 beam, and two plates hold everything together.

Figure 4–10:
Simple Platform

An alternative is to eliminate the plates and use double-pegs to hold the beams together. (The name "double-peg" is actually somewhat of a misnomer since it is only 1.5 times as long as a normal peg.)

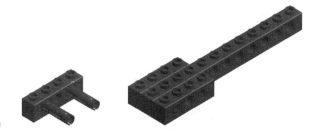

Figure 4–11:
Wide Beam Platform

This technique of attaching beams together with pegs often comes in handy when building motors into structures that are just a little too wide. Beams provide much better structural support than plates, so the motor is less likely to pop off your model if it starts or stops suddenly.

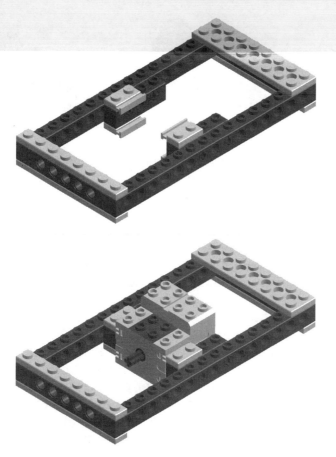

Figure 4–12:
Using Wide Beams to Mount a Motor

BASIC GEARS

Motors supply the mechanical power to bring a robot alive, but in many cases the motor itself may be in the wrong position, turning at the wrong speed, or providing the wrong amount of torque. The following pages discuss how gears can be used to transfer mechanical power and to convert speed into torque.

Gear Reduction

Consider what happens as the pulley with the handle is turned clockwise in the model below. The pulley is on the same axle as a small gear, which then meshes with a larger gear, which in turn is attached to a second pulley. The numbers in the illustration indicate the length of the axle.

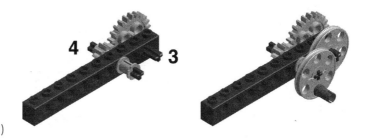

Figure 4–13:
Gear Reduction (3:1)

You may have noticed that the second pulley rotates differently than the first one. First of all, it is spinning in the opposite direction. This is one of the basic properties of the way gears mesh with one another—the direction of rotation is reversed.

The second observation is that the second pulley turns much more slowly than the first one. If you count, you'll find that the first pulley must make 3 complete revolutions for the second pulley to turn once. This difference in speed occurs because the two gears have different numbers of teeth. If you feel like counting, you'll find that the small gear has 8 teeth and the larger one has 24. Thus, in order for the larger gear to make one complete revolution, all 24 teeth must move past the point where the gears mesh. Since the small gear only moves 8 teeth past this point per revolution, it must turn 3 times (24 divided by 8) for each revolution of the larger gear. Since the number of teeth on a gear determines how it meshes with other gears, it is worth remembering that the four standard LEGO gears have 8, 16, 24, and 40 teeth, respectively.

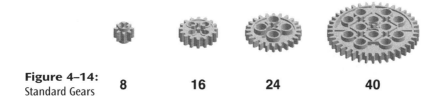

Figure 4–14:
Standard Gears

8   16   24   40

Reducing the number of revolutions is called *gear reduction,* or *gearing down.* The exact amount of gearing down can be written as the ratio that specifies how many turns of the first gear equal a certain number of turns of the second gear. In the above example, the gear ratio is 3:1.

Speed and Torque

In the previous example we used gears to reduce the speed of rotation. We can also arrange things so that we increase speed. This is called *gearing up,* and it can be accomplished by having a large gear turn a smaller one. The example below uses a 40-tooth gear to turn an 8-tooth gear; it thus has a gear ratio of 8:40 = 1:5 (you can reduce gear ratios just like you reduce fractions).

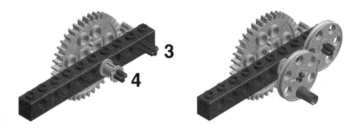

Figure 4–15:
Gearing Up (1:5)

If you start turning the pulley with the handle, you'll find that the second pulley spins much faster (5 times as fast to be exact). All of this extra speed seems like we're getting something for nothing. Unfortunately, physics gets in the way and prevents us from having a free lunch.

A rotating axle has both speed and torque. Speed represents how fast the axle is spinning, and it is typically measured in units such as revolutions per minute (rpm). Roughly speaking, torque represents how much force the rotating axle is capable of supplying. Higher torque means that the axle will be able to propel a heavier vehicle or lift a heavier weight.

> Technically, torque is a product of force and distance. If an axle has torque of 1 ft.-lb., it means that if a wheel of a radius of 1 ft. were attached to the axle, there would be 1 pound of force at the edge of the wheel. For a given force, the amount of torque increases as the radius of the rotating object increases. That's why screwdrivers with fat handles are better at loosening tight screws.

When gears are used to change the speed of an axle (gearing up or down), they also affect the torque in the opposite manner. For example, if you gear down using a 2:1 ratio, the speed will be cut in half but the torque will be doubled. Conversely, when you increase speed, you reduce torque. In the 1:5 example above, the torque was reduced by a factor of 5. The total amount of work that can be done (speed times torque) remains unaffected by gear ratios (aside from the small amount of energy that is wasted due to friction whenever two gears mesh).

Compound Gear Trains

Our previous examples used only two gears. It is also possible to combine multiple stages of gears to form a compound gear train. When compounded, the ratios of the individual stages are multiplied together to determine the overall gear ratio. This allows the creation of gear ratios far larger (or smaller) than those available from a single pair of gears.

The example below is a three-stage gear reduction, with each stage consisting of an 8-tooth and a 24-tooth gear. Note how the end of one stage (24-tooth gear) is attached to the start of the next stage (8-tooth gear). Each stage has a ratio of 3:1. The ratio of the compound arrangement is 3:1 x 3:1 x 3:1 = 27:1 (ratios multiply just like fractions).

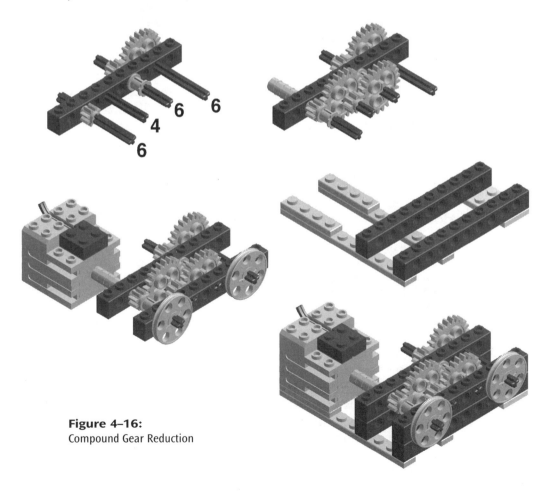

Figure 4–16:
Compound Gear Reduction

We can use the RCX to power the motor. First you'll need to download the program from chapter 1 to the RCX. Attach the motor to RCX output A and run the program. The motor should start spinning quite fast (as you can see by the left pulley), while the right pulley at the end of the gears moves very slowly. By grasping the left pulley you should be able to prevent the motor from turning. Now try stopping the motor by grasping the right pulley. In the first case it was fairly easy to overcome the motor's torque. In the second, we have to deal with 27 times more torque.

Distances between Gears

Some combinations of gears, such as a 24-tooth and an 8-tooth, line up perfectly along a beam. For other combinations, such as a 16-tooth and a 24-tooth gear, the spacing of holes in a beam just won't work.

So far we have been describing gears by the number of teeth they have. Gears also have another important measurement: their radius. When you are meshing two gears, the distance between their axles should be equal to the sum of their radii. The radii (in LEGO units) for the four standard gears are shown in the table below:

Teeth	Radius
8	0.5
16	1.0
24	1.5
40	2.5

Table 4–1

Now it should become clear why the 16/24 combination doesn't work—we need a spacing of 2.5 units, and the holes in a beam have whole-number spacing (no pun intended). Note that the 8-, 24-, and 40-tooth gears can mesh with each other along a beam, and the 16-tooth gear can mesh with itself. It is only when combining the 16-tooth with one of the others that we have difficulty.

With a little cleverness, however, we can get the gears to mesh. The example below shows how to use two *crossblocks* separated by a *half bushing* to achieve the proper spacing for an 8-tooth gear and a 16-tooth gear.

Figure 4–17:
Nonstandard Spacing

A more flexible solution is to use vertical and horizontal distance to create a diagonal spacing that works. To compute the diagonal distance, you use the Pythagorean Theorem. There is a slight complication in that a brick is actually 1.2 units tall (see the earlier section on structures for more discussion of this).

$$distance = \sqrt{width^2 + (1.2 \cdot height)^2}$$

In the example below the gears are separated horizontally by 2 units, and vertically by 1 brick and 1 plate. The resulting distance is just a touch over 2.5 units, but it is still close enough to mesh 16- and 24- tooth gears together.

Figure 4–18:
Diagonal Spacing

$$distance = \sqrt{width^2 + (1.2 \cdot height)^2} = \sqrt{2^2 + (1.2 \cdot (1 + \tfrac{1}{3}))^2} \approx 2.56$$

SPECIALIZED GEARS

Besides the "vanilla" gears discussed above, specialized gears such as the *crown gear*, the *bevel gear*, the *worm gear*, the *rack*, and the *differential* are also available. The following pages describe the uses for each of these special gears.

Crown and Bevel Gears

The *crown gear* is the same size as a standard 24-tooth gear, but its teeth are curved to allow it to mesh with a gear at right angles to it. The meshing involves a lot of friction, so this is not a good arrangement for something where speed or power is at a premium. However, for auxiliary functions the crown gear can be quite useful. It is also important that the crown gear have ample support directly behind the meshing point; otherwise it will tend to slip.

Figure 4–19:
Using a Crown Gear

The *bevel gear* is a more compact alternative to the crown gear. Its 12 teeth can only mesh with other bevel gears, but when used properly it can transfer power a bit more efficiently than the crown gear. It is critical that the bevel gear be backed up against something solid, preferably a beam, although bushings can be used to fill empty space.

Figure 4–20:
Using Bevel Gears

Worm Gear

The *worm gear* is unique. Each full rotation of the worm gear results in the movement of one tooth of the gear it is meshed with. In the example below, it is meshed with a 24-tooth gear, and the resulting gear ratio is thus 24:1. Needless to say, the worm gear provides an extremely compact mechanism for gear reduction.

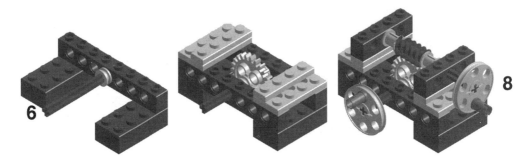

Figure 4–21:
Using the Worm Gear

The worm gear also provides a one-way mechanism. Try turning the left pulley shown in the above example. There will be some wiggle (due to the fact that the worm gear is just a little shorter than 2 units), but no amount of force will cause the worm gear itself to rotate. This means that the worm gear can transfer power from its axle to a gear meshed with it but will not allow power to be transferred in the reverse direction.

This feature can be exploited when building something like a crane, where a motor is used to lift some heavy weight and the weight must remain lifted even after the motor is turned off.

The wiggle mentioned above is actually one of the more troublesome aspects of using the worm gear. In some cases, a little wiggle is acceptable, but in other cases it needs to be eliminated. One option is the vertical arrangement shown below. If such an arrangement is used, then suitable vertical bracing must be used; otherwise the vertical forces on the worm gear tend to pop the structure apart.

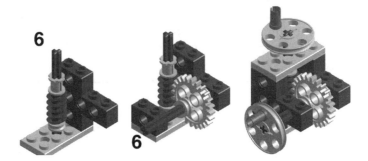

Figure 4–22:
Using the Worm Gear Vertically

Multiple worm gears may be placed end to end to create a longer worm gear. Care must be taken, however, to align the gears. If they are twisted with respect to one another, then the meshing will be uneven at the seam between the two worm gears.

Figure 4–23:
Multiple Worm Gears

Right   **Wrong**

The Rack

A *rack* can be used to convert the rotation of a gear into linear movement. It is common to use an 8-tooth gear in combination with a rack. Larger gears will still mesh properly, but in most applications the finer control afforded by the smaller gear makes it the best choice.

The rack needs to be able to slide freely back and forth; hence it must rest on a smooth surface. Specifically, you don't want it sitting on top of a normal brick or plate. LEGO makes some smooth-topped pieces (called tiles) that work well for this; however, the MINDSTORMS Robotics Invention System doesn't include any tiles. A little creativity will solve the problem, though. In the model below, we use two *half-beams* to create a smooth surface for the rack. Turning the crank will cause the rack to move back and forth.

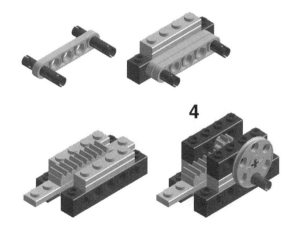

Figure 4–24:
Using a Rack

The rack itself is only 4 units long, so it doesn't take much turning for it to be pushed completely out of the mechanism. A longer rack can be built by placing multiple racks end to end.

The Differential

The *differential* is the most complicated gear assembly, but it is also one of the most useful. Internally, it uses three bevel gears to connect the two independent axles with the shell of the differential. The axles may rotate at different speeds, but their average speed must equal the rotation of the differential shell. Assembling the differential can be a bit tricky. The "middle" bevel gear (the one not attached to an axle) must be placed inside the differential shell first. Then each axle may be inserted through the differential to its own bevel gear. The completed assembly is shown below.

Figure 4–25:
The Differential

The best way to understand the differential's operation is to play a little; start by building this model:

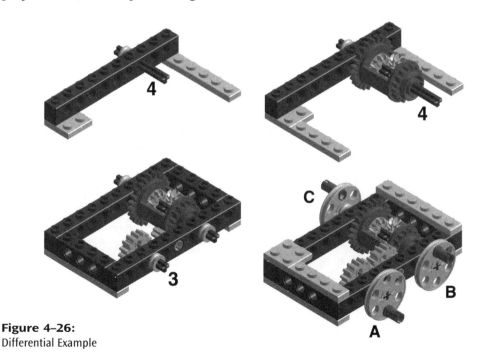

Figure 4–26:
Differential Example

Pulley A turns the shell of the differential; pulleys B and C are connected to the differential's independent axles. Turning pulley A one full rotation causes the differential to also make one full rotation. This results in pulleys B and C also making one full rotation (assuming neither one has too much friction). This is the simplest case: everything turning at the same speed.

If you turn pulleys B and C at the same speed but in opposite directions, the differential (and pulley A) will remain motionless. In this case, the average speed of the independent axles is zero, so the differential remains still.

Play with other combinations until you feel you are comfortable with how the differential works. What happens if you turn the differential but keep one of the other axles frozen in place?

PULLEYS AND BELTS

Like gears, *pulleys* can be used to convert between speed and power, but the similarity ends there. Whereas gears reverse direction when they mesh, pulleys turn in the same direction. And, unlike gears, which have teeth to be counted, the ratio between pulleys is determined strictly by their size. While you should never rely on pulleys for exact ratios (due to slippage), their relative sizes are shown below:

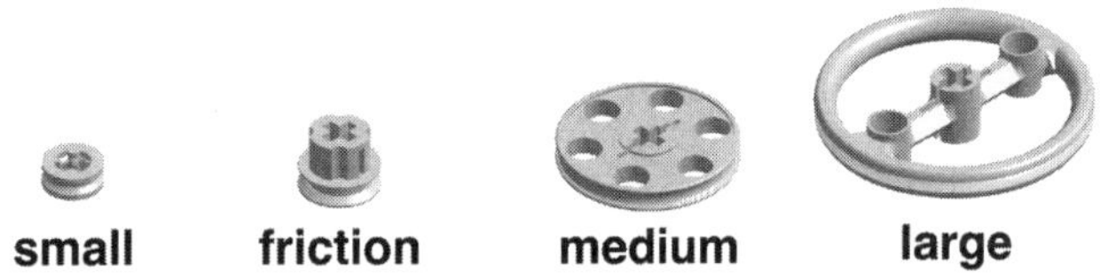

Figure 4–27:
Pulleys

| small | friction | medium | large |

Pulley	Relative Size
small	2
friction	3
medium	7
large	11

Pulleys do not mesh with each other directly; power is transferred from one pulley to another via a rubber *belt*. This means that the distance between pulleys is a little more flexible than the distance between gears. In some cases this allows pulleys to be used in places where a combination of gears wouldn't fit. In the model below, it takes roughly 3½ turns of the crank for the larger pulley to complete one revolution.

Figure 4–28:
Pulley Example

Now try holding the second axle (the one with the larger pulley) still while turning the crank. The belt will start to slip. The tightness of the belt determines how much power can be transferred before the belt starts to slip. At first thought, such slipping may seem to be a limitation, but it is actually a very important feature of pulleys that can be exploited in designs. It can provide a sort of safety valve in systems where a motor is powering something with limited range of motion.

For example, let's try to motorize the rack mechanism from the previous section. Instead of a crank, we use a 24-tooth gear. We also build "stoppers" to prevent the rack from moving too far. Finally, we add a motor which drives the track in a 3:1 ratio. Use the RCX program from chapter 1 to power the motor and test the motorized rack. When I tried it, my model broke apart; the force of the motor was too much for the stoppers to handle.

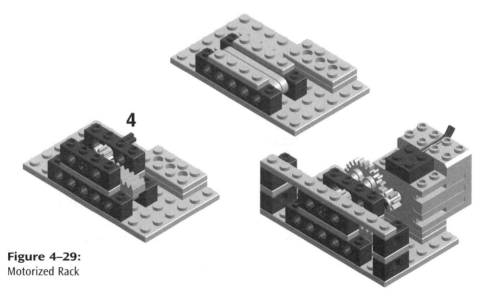

Figure 4–29:
Motorized Rack

We could reinforce the stoppers, but then something else would probably break. The forces from the motor will just find the weakest link and break it. Perhaps the beams will separate, or the motor itself will pop off. Reinforcing every single weak point in a large and complex model can be quite an undertaking.

Pulleys and belts provide us with a simpler solution. We can replace the 24-tooth gear with a medium pulley and replace the 8-tooth gear on the motor with a small pulley. A blue belt completes our model.

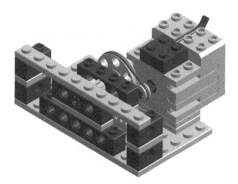

Figure 4–30:
Motorized Rack with Pulleys

Not only is the pulley-and-belt solution simpler than reinforcing the entire model; it is also a better solution. If you reinforce a model to the point where the forces from the motor will not be able to break it, then the motor will be prevented from turning once the rack hits a stopper. This is known as *stalling* the motor, and it tends to put more strain on the motor than normal operation. Ironically, a motor also consumes more energy when it is stalled than when it is running. This means that a stalled motor will also drain your batteries faster.

RATCHETS

A ratchet is a mechanism that freely allows rotation in one direction but prevents rotation in the opposite direction. There aren't any special LEGO ratchet pieces, but we can fashion our own ratchets using the pieces at hand.

A Simple Ratchet
Our first example is a simple ratchet mechanism built from a gear and a beam. As the pulley is turned clockwise, the teeth on the gear hit the arm in such a way that it is briefly lifted out of the way.

As the arm repeatedly lifts and falls into place, you can hear the characteristic clicking sound of a ratchet mechanism. Now try turning the pulley counterclockwise. Instead of lifting the arm out of the way, the gear's teeth actually push it more firmly into place, thus preventing movement.

Figure 4–31:
A Simple Ratchet

Finally, try giving the pulley a very quick clockwise spin. If you spin it fast enough, it will kick the arm hard enough that it swings completely behind the gear, rendering the ratchet ineffective. For this reason, it is sometimes desirable to build some sort of stop to prevent the arm from swinging too far out of position.

Our simple ratchet relies on gravity to hold the arm in place. Sometimes the ratchet may not be able to count on gravity (perhaps it needs to operate upside down). In these cases, another mechanism, such as a rubber band, must be provided to keep the arm in place. However, a strong rubber band will require the gear to supply more force when lifting the arm out of the way, thus sapping some of the power that is presumably destined for some other function. Like many other techniques, building a good ratchet is a balancing act between efficiency and robustness.

The Ratchet Splitter

Sometimes a robot needs to do more than three things. Since the RCX has only three outputs, it is necessary to find a way for multiple functions to share the same motor (and hence the same RCX output). The *ratchet splitter* provides a way for a single motor to control two separate functions. The catch is that each function can only operate in a single direction, and only one function may operate at a time. The design is relatively simple, relying on a differential and two ratchets.

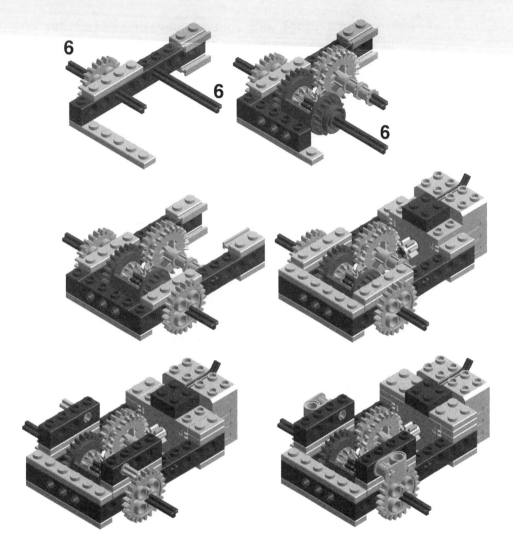

Figure 4–32:
The Ratchet Splitter

As discussed before, a ratchet allows rotation in one direction but prevents the reverse motion. The two ratchets in this model are set up so that one axle may rotate only clockwise, while the other may rotate only counterclockwise. Power is then applied to the shell of the differential. Consider what happens when the motor turns the differential in the forward direction: the "normal" operation would be for both axles to rotate forward at the same speed as the differential, but one axle is prevented from going in this direction (by the ratchet); therefore the other axle must spin twice as fast (so the average axle speed equals that of the differential shell). If the motor is reversed, the previously spinning axle is held in place by its ratchet while the other axle is free to rotate (again at twice the speed).

Levers

The lever is one of the most basic machines. With it, a small force can be used to lift a large weight. Consider the example below. Placing one brick on each side will result in the long beam being balanced.

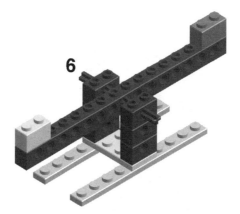

Figure 4–33:
A Balanced Lever

Placing the bricks closer together still leaves the beam balanced.

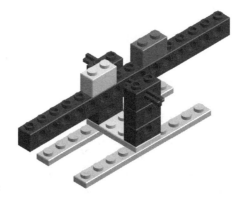

Figure 4–34:
Another Balanced Lever

Now if we place two bricks on the left side while leaving only one brick on the right, the beam no longer balances—it tips completely to the left. This should come as no surprise because two bricks obviously weigh more than a single brick.

Figure 4–35:
Two Bricks Are Heavier Than One

In the figure below, however, things have changed. There are still two bricks on the left and a single brick on the right, but somehow the right side is "heavier" than the left.

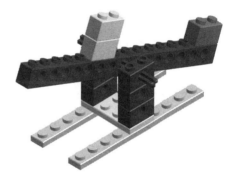

Figure 4–36:
Making One Brick "Heavier" Than Two

In our example, the beam is actually a lever. The point at which the axle goes through the beam is called the *fulcrum* of the lever. When a force is applied to a lever, both the magnitude of the force and the distance between the force and the fulcrum must be taken into account. The product of the force and distance is called the *moment.* In our example, the beam will tip to the side with the greater moment, or balance if the moments are equal. In the first three cases, the distances for the left and right bricks were equal, so only the number of bricks (and hence the weight on the beam) mattered. In the example above, the distances are not equal. In this case, the left side has two bricks that are 2 units away from the fulcrum, which equals a moment of 4. The right side has only 1 brick, but it is 7 units away, for a moment of 7. Since 7 is greater than 4, the beam tips to the right. If we put the right-hand brick 4 units away from the fulcrum, then its moment will be the same as that of the left-hand side, and the beam will once again balance, as shown below.

Figure 4–37:
Balancing Bricks

All of our examples have shown the fulcrum in the middle, with the forces applied on either side, but this does not have to be the case. The fulcrum can be at one end, and the forces can both be applied to the same side. Bottle openers, the claw end of a hammer, and crowbars are all examples using a lever so that a modest force at one end can create a much larger force nearer to the fulcrum.

CONCLUSION

This chapter has presented a lot of different topics: frames, gears, pulleys, levers, differentials, vertical bracing, levers, and ratchets. Don't worry if it seems like a lot to remember; practice makes perfect, and the remaining chapters of the book will provide plenty of practice. This concludes the Fundamentals section of the book, and we are now ready to start building some robots.

PART TWO

Robots

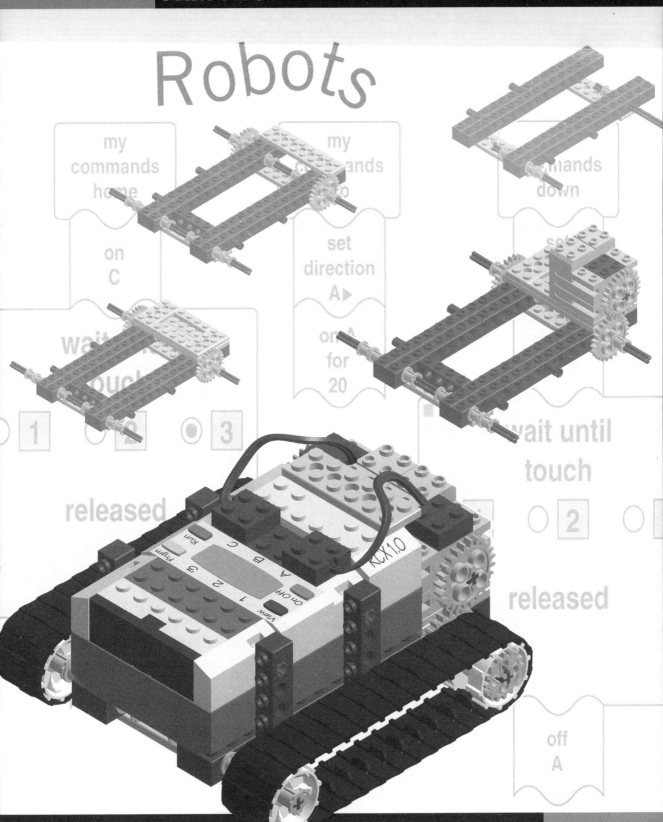

Chapter 5
Tankbot

Tankbot is a rather simple robot; in fact calling it a robot may be a bit of a stretch. Tankbot has no sensors, so it cannot respond to the environment. All it can do is follow a preprogrammed set of directions (e.g., drive straight for 2 seconds, turn right, etc.). However, Tankbot itself will become the base for some more complicated robots presented in later chapters.

Like a tank, Tankbot uses caterpillar treads instead of wheels to propel itself (this is also known as tracked vehicle). Caterpillar treads, hereafter referred to simply as treads, are often a good choice for general purpose robotic vehicles. They are easy to construct and program and have the ability to rotate a vehicle in place.

CONSTRUCTION

Construction of Tankbot begins with a frame to support the RCX and motors. In Step 1, beams are prepared with a number of friction pegs. These pegs will be used to join other beams together in later steps.

Figure 5-1:
Tankbot Step 1

In Step 2, the frame is completed with an additional set of beams and a pair of 2 x 8 Technic plates. The reason for doubling up the beams is that the RCX is 8 units wide, but because of its tapered design, it needs to be supported somewhere other than its edges. The doubled-up beams provide a frame that is as wide as the RCX while still providing interior support. 2 x 8 plates were chosen instead of 1 x 8 plates in order to make the frame more rigid. Two axles, which will eventually connect to one tread, are also put in place. A similar pair of axles for the other tread are added in Step 3. Note that the axles are 12 units apart—this is the ideal spacing for the treads that are to be used on Tankbot.

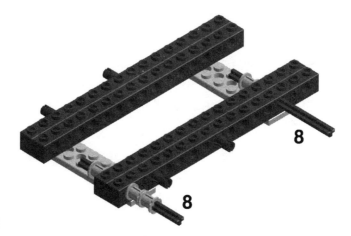

Figure 5-2:
Tankbot Step 2

The rear axles have 24-tooth gears, which will later mesh with gears on the motors. On all four axles, bushings are used inside to lock the axles in place and to provide adequate spacing for the tread hubs (to be added later).

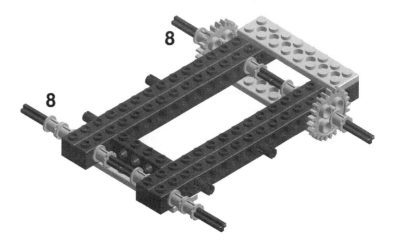

Figure 5-3:
Tankbot Step 3

In Steps 4 and 5, a platform to support the motors is constructed. The motors do not have a flat bottom, so sometimes rather unusual constructions must be used to support them. In this case, the recessed center area provides space for the rounded bottom of the motors.

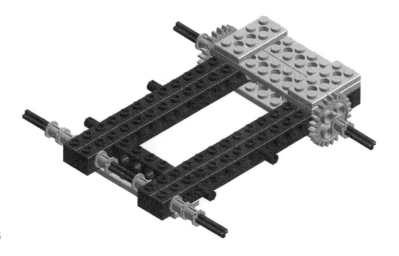

Figure 5-4:
Tankbot Step 4

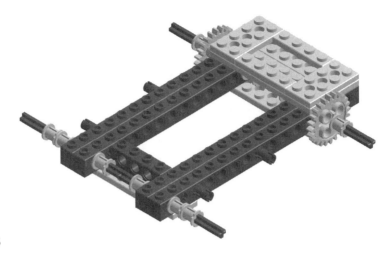

Figure 5-5:
Tankbot Step 5

The motors themselves are added in Steps 6 and 7, and the RCX is added in Step 8. It is important to connect the wires exactly as shown. This will ensure that when the RCX activates an output in the forward direction, the motors will spin in the proper direction to propel Tankbot forward. For consistency with some of the later examples, we will consider the motors to be at the back of Tankbot and the IR window of the RCX to be at the front.

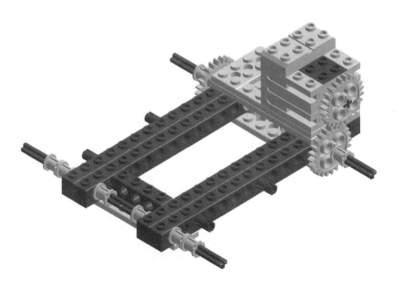

Figure 5-6:
Tankbot Step 6

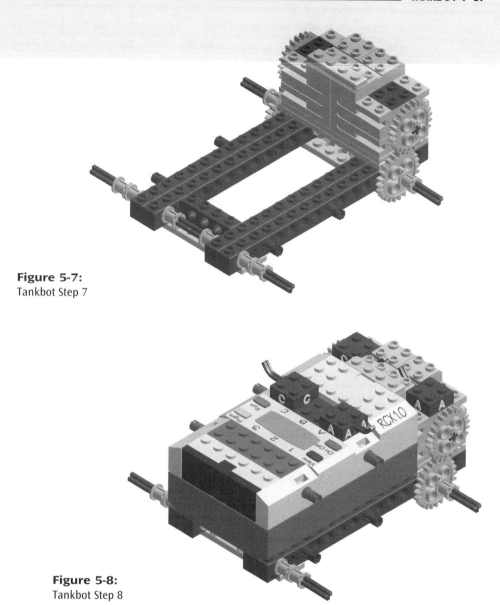

Figure 5-7:
Tankbot Step 7

Figure 5-8:
Tankbot Step 8

Step 9 is the final step for Tankbot. Four 1 x 6 beams are used to secure the RCX to the frame. This is actually quite important since it is tempting to pick up Tankbot by grasping the RCX itself. Without the 1 x 6 beams, the top of the RCX could then separate from the bottom—often dumping the batteries on the ground.

The easiest way to attach a tread is to wrap the tread around the hubs, then slide the tread and both hubs onto the axles. This is far easier than attempting to stretch the tread around hubs that are already in place. The hubs themselves slide and spin freely on axles. This makes it necessary to secure them in place. In the case of the rear treads, 16-tooth gears must be used. This ensures that the hub will turn whenever the axle turns. For the front hubs, it is not necessary to lock them to the axle; simple bushings are sufficient to hold them in place (in fact, using gears on the front axles would lead to increased friction and actually degrade overall performance).

Figure 5-9:
Tankbot Step 9

DRIVING STRAIGHT

Our first program is very simple—all it does is turn on both motors so that Tankbot travels in a forward direction. Versions of this program in both RCX Code and NQC are shown.

Figure 5-10:
Tankbot Program 1 in RCX Code

```
// tankbot1.nqc - drive straight ahead

#define LEFT OUT_A
#define RIGHT OUT_C

task main()
{
```

```
    On(LEFT+RIGHT);
    until(false);
}
```

Download the program to the RCX and try running it. Does Tankbot start driving straight ahead? If Tankbot moves backwards, or spins with one tread moving forward and one tread backwards, then you need to change the wiring so that the RCX's notion of "forward" matches reality.

In some cases Tankbot will curve slightly to one side or the other as it is driving. This is quite normal and is a consequence of the slight differences between the individual motors, treads, and axles. These slight variations can cause one side to encounter a little more friction than the other, thus turn at a slightly slower speed. In later chapters, we will discuss other drive mechanisms that avoid this problem, but for now we'll just have to live with it.

If you are new to using NQC, then a few comments about the code are probably in order. First of all, note how the first two lines define the symbols LEFT and RIGHT to be OUT_A and OUT_C, respectively. This isn't strictly necessary, since without those definitions, we could just use

```
On(OUT_A + OUT_C);
```

However, it is a good programming practice to assign names to each output and sensor near the top of the program. Within the body of the program, they can be referred to by function (e.g., LEFT) rather than having to remember which output or sensor was being used. This makes the program more readable, and allows it to be adapted easily if you change how the motors or sensors are connected later on.

The second detail about the code is the last statement:

```
until(false);
```

Without this statement, the program would finish running right after turning on the motors. The motors would then be left running, and the only way to stop them would be by turning the RCX off. The RCX does allow you to stop a running program by pressing the **Run** button, and this also has the side effect of turning off all of the motors. However, since the program would already be finished, pressing **Run** couldn't stop it—instead it would start running the program a second time! Using the until statement causes the program to enter an infinite loop (waiting for something that will never be true), thus the program runs forever. This allows us to use the **Run** key to stop the program and the motors.

TURNING

Most of the wheeled vehicles we encounter in real life, such as bicycles and automobiles, can be steered by changing the direction that some of the wheels face, thereby directing the entire vehicle to start curving left or right. Tankbot, however, has no means of changing the angle of its treads. How can we steer it?

The answer comes from geometry. Consider a car driving in a circle. If we were to look at the tracks left by the car's wheels we would see that they form two concentric circles; the outer wheels traveling in a circle that is slightly larger than the inner wheels. Since both sets of wheels take the same amount of time to complete the circle, the outer wheels must cover more ground per second. Assuming that the wheels are all the same size, this means the outer wheels are turning slightly faster than the inner wheels.

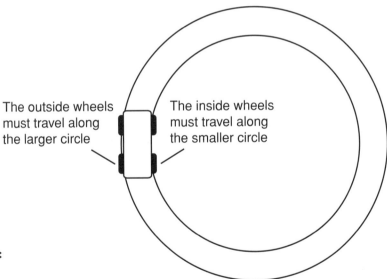

The outside wheels must travel along the larger circle

The inside wheels must travel along the smaller circle

Figure 5-11:
Turning

We can apply this fact in reverse to steer Tankbot. We simply run one side's tread faster than the other side and Tankbot will start turning.

We'll start with a program that causes Tankbot to weave back and forth. We can do this by repeating the following steps:

- turn right

- drive straight

- turn left

- drive straight

Turning is accomplished by running one motor at half the power while the other one is at full power.

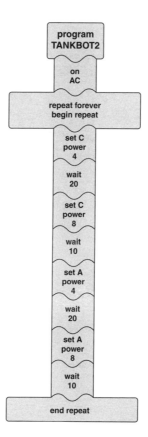

Figure 5-12:
Tankbot Program 2 in RCX Code

```
// tankbot2.nqc - drive and turn

// motors
#define LEFT  OUT_A
#define RIGHT OUT_C

// how much time to spend turning or forward
#define TURN_TIME          200
#define STRAIGHT_TIME      100

// speed to run a turned motor
#define TURN_POWER    3

task main()
{
    // start with both motors on
    On(LEFT+RIGHT);

    // repeat the following steps forever
    while(true)
    {
        // turn right by slowing down the right tread
        SetPower(RIGHT, TURN_POWER);
        Wait(TURN_TIME);

        // resume going straight
        SetPower(RIGHT, OUT_FULL);
        Wait(STRAIGHT_TIME);

        // turn left
        SetPower(LEFT, TURN_POWER);
        Wait(TURN_TIME);

        // resume going straight
        SetPower(LEFT, OUT_FULL);
        Wait(STRAIGHT_TIME);
    }
}
```

As usual, we define anything that we may want to change later (such as the output ports, the time spent turning, and the speed to use for turns) as constants near the top of the NQC program.

The results were probably less than spectacular. The amount of turning may have been barely noticeable, and the sharpness of the right turns was probably different from that of the left turns. Why doesn't this work?

The RCX cannot directly control the speed of a motor. Instead, it varies the amount of power supplied to the motor. The LEGO motors were designed to be very power efficient, and they have an internal flywheel which acts as a sort of energy storage tank. The flywheel is most effective when the motor has very little physical resistance (called load). Starting the motor consumes a lot of energy since the flywheel must also be spun up to speed. Once running, however, a lightly loaded motor (perhaps a motor that is turning a gear that isn't connected to anything else) can maintain its speed with very little external power. In this case, it doesn't matter if the RCX is sending lots of power or just a little bit to the motor, it will turn at the same speed regardless. Under a heavy load (propelling a vehicle across a carpeted surface), the flywheel is less effective, and changing power levels will have a more noticeable effect on speed.

With Tankbot, we could try adjusting the power level used for turns (by setting TURN_POWER), or even turn a motor completely off. However, one of the unique features of a tread-based design is that it is quite good at turning in place by reversing one tread while leaving the other running. In fact, it is so good that we can reduce the amount of time spent turning from 2 seconds to $1/2$ second and thereby increase the time spent driving forward.

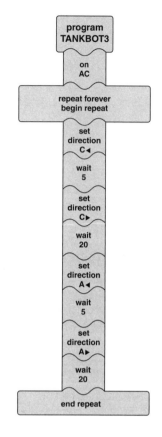

Figure 5-13:
Tankbot Program 3 in RCX Code

```
// tankbot3.nqc - improved turning

// motors
#define LEFT  OUT_A
#define RIGHT OUT_C

// how much time to spend turning or forward
#define TURN_TIME        50
#define STRAIGHT_TIME    200

task main()
{
    // start with both motors on
    On(LEFT+RIGHT);
```

```
// repeat the following steps forever
while(true)
{
    // turn right by reversing the right tread
    Rev(RIGHT);
    Wait(TURN_TIME);

    // resume going straight
    Fwd(RIGHT);
    Wait(STRAIGHT_TIME);

    // turn left by reversing the left tread
    Rev(LEFT);
    Wait(TURN_TIME);

    // resume going straight
    Fwd(LEFT);
    Wait(STRAIGHT_TIME);
}
}
```

CONCLUSION

Congratulations, you have completed the first robot in this book. Although relatively simple, the construction and programming techniques used in Tankbot serve as a foundation for many of the other robots in this book.

Tankbot itself can be adapted for a variety of tasks. In the next few chapters, a variety of sensors will be added to Tankbot. Bumpbot and Bugbot use touch sensors to navigate around obstacles. Linebot uses a light sensor to follow a line drawn on the floor. Scanbot uses a light sensor (and an extra motor) to look around and seek out bright lights.

Chapter 6

Bumpbot

The goal of Bumpbot is to drive around randomly without getting stuck against a wall or in a corner. To accomplish this, Bumpbot needs some way of detecting a collision with an obstacle. This can be achieved by adding a "bumper" to the basic Tankbot chassis. This chapter presents two different versions of Bumpbot, each with its own bumper design.

A SIMPLE APPROACH

A simple bumper can be constructed from a touch sensor as shown below. A #12 axle is used as the bumper itself, which is attached to a #6 axle that presses against a touch sensor. We mount this sensor on the front of our previous robot (Tankbot), and presto— Bumpbot is born.

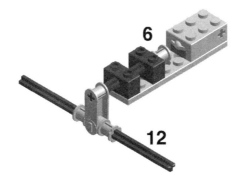

Figure 6–1:
A Simple Bumper

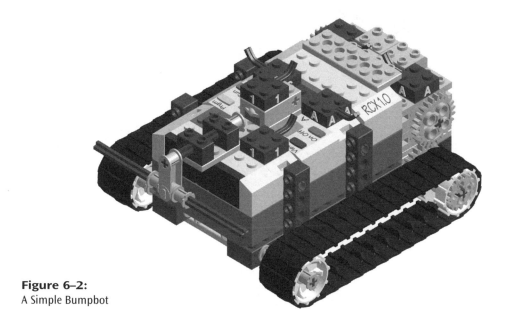

Figure 6–2:
A Simple Bumpbot

All that remains is a little programming. Let's start by pro-gramming Bumpbot to drive forward until it hits something, then back up a little, spin around a bit, and resume going forward. A suitable program in RCX Code is shown below:

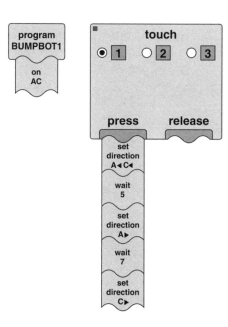

Figure 6–3:
Simple Bumpbot Program in RCX Code

A similar program can also be created in NQC.

```
// bumpbot1.nqc - a simple bumper

// sensors
#define BUMP SENSOR_1

// motors
#define LEFT  OUT_A
#define RIGHT OUT_C

// constants
#define REV_TIME      50
#define SPIN_TIME     70

task main()
{
    // configure the sensor
    SetSensor(BUMP, SENSOR_TOUCH);

    // start going forward
    On(LEFT+RIGHT);

    // do this forever
    while(true)
    {
        // wait for bumper to hit something
        until(BUMP == 1);

        // back up
        Rev(LEFT+RIGHT);
        Wait(REV_TIME);

        // spin around
        Fwd(LEFT);
        Wait(SPIN_TIME);

        // resume
        Fwd(RIGHT);
    }
}
```

Because Bumpbot uses a sensor, there are a few new elements to our NQC program. The first thing we must do whenever sensors are being used is to tell the RCX what type of sensor is connected to the sensor port. This is done with the SetSensor command, which takes two arguments: a sensor port and a sensor type. In the above example, BUMP was simply defined as another name for SENSOR_1—just like we've been defining LEFT and RIGHT to be names for OUT_A and OUT_C.

We've used the until statement before, but this time the sensor is used as part of the condition. A task will stay in the until statement until its condition becomes true. In this case, we are waiting for BUMP == 1. BUMP is just another name for the touch sensor, meaning that we are waiting for the touch sensor's value to be equal to 1.

> NQC inherits much of its syntax—both good and bad—from the C language. In C, = and == mean very different things. Specifically, = is used to assign a value to a variable, whereas == is used to test equality (compare two values). This is often confusing to those new to the language. There's good reason for such a distinction in C, but less so in NQC. Nevertheless, NQC sticks with the C syntax.

A BETTER BUMPER

Our first bumper is very simple to build, but it does have a problem in that anytime it hits something, the force of impact is transmitted directly to the touch sensor and, hence, to the rest of the model. In addition, the RCX cannot respond instantaneously, so for a short time the motors keep running, resulting in even more pressure being applied to the bumper and sensor. The most noticeable effect of this is that after enough collisions, the bumper will just fall off.

Although we could mount the bumper more securely (perhaps using some vertical bracing), this still wouldn't eliminate the stress to the touch sensor. Such repeated impact is something to be avoided, so what we really need is a bumper that will give way during impact and allow more time for the RCX to respond.

We begin constructing the improved Bumpbot by removing the treads and front 1x6 beams from Tankbot as shown in Step 1.

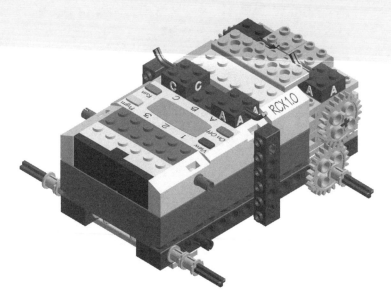

Figure 6–4:
Bumpbot Step 1

Supports for the bumper assembly are then constructed in Steps 2 and 3. The bumper itself is going to be a pivoting structure held in the "forward" position by a rubber band. The gray peg at the front of the 1x16 beam serves as the pivot point, while the black peg will serve as an anchor point for the rubber band. Although these pegs look alike (except for color), they have slight differences, which alter how they are used. The gray peg is slightly smaller than the holes in the beams (and other parts), and as a result parts joined with this peg can pivot freely. In contrast, the black peg is slightly larger and holds parts together more firmly. For that reason, the black peg is often called a *friction peg,* and should be used whenever pivoting is not required.

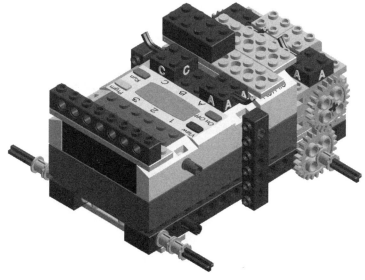

Figure 6–5:
Bumpbot Step 2

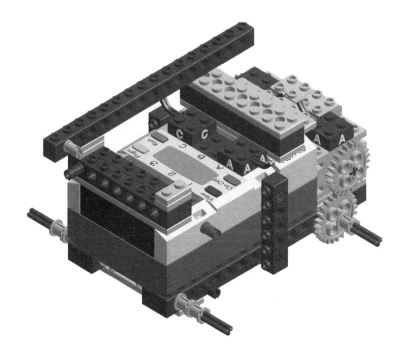

Figure 6–6:
Bumpbot Step 3

Step 4 shows the construction of one half of the bumper. The black *angle beam,* along with an axle, makes for a nice bumper, and the yellow *double-angle beam* provides a convenient way to extend the bumper in front of the RCX. Both of these parts are often good choices when designing bumpers. The black angle beam is attached to the axle using a special piece called a *catch* (see Step 6 for a better view of this).

Figure 6–7:
Bumpbot Step 4

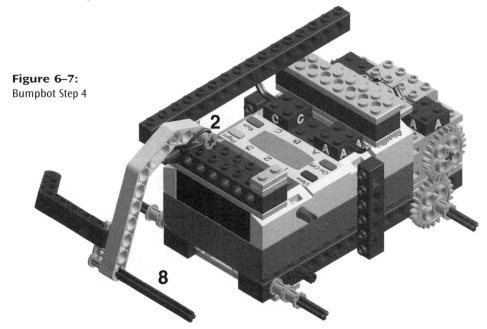

One end of the double-angle beam has a bushing and #2 axle attached to it. These parts serve a dual purpose: they provide the other anchor point for the small black rubber band, and they also provide the means for the bumper to activate the touch sensor, which is added in Step 5. The sensor should be connected to port 1 on the RCX. The extra 1x2 plates on top of the touch sensor have an important function, which will become clear a little later.

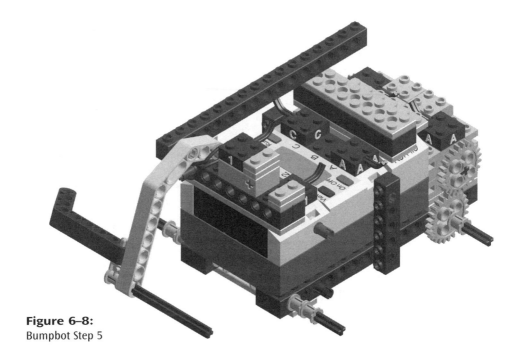

Figure 6–8:
Bumpbot Step 5

The other half of the bumper is completed in Step 6. This half is almost identical to the first half, with the exception that it does not need the #2 axle and bushing.

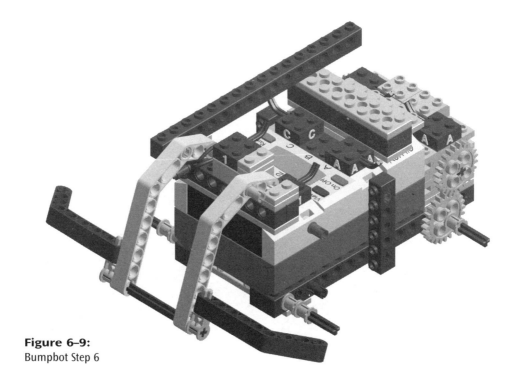

Figure 6–9:
Bumpbot Step 6

Step 7 finishes off the robot. Tankbot used four 1x6 beams to secure the RCX to the bottom frame. In Bumpbot, a pair of 1x6 beams is still used to secure the back of the RCX, while the front is braced using 1x10 beams that go from the bottom frame all the way up to the 1x16 beams supporting the bumper itself. This vertical bracing keeps the bumper intact even after multiple collisions. As a final detail, the treads and hubs must be attached and locked into place (once again with bushings on the front and 16-tooth gears in back).

Figure 6–10:
Bumpbot Step 7

This new bumper design has a larger range of motion than the simple bumper presented at the start of the chapter. The bumper is normally held forward by the tension of the rubber band. In this position, the touch sensor is pressed by the small arm at the end of the double-angle beam. During a collision, the front of the bumper gets pushed in, leading to a rotation of the double-angle beam and lifting the arm away from the touch sensor. After the touch sensor is released, the bumper can continue to be pushed in a little further without any serious impact to Bumpbot itself. This gives the RCX extra time to react to the sensor. Even in the worst case, any impact would be felt by the structure of the bumper and not by the touch sensor itself.

Remember those extra plates on top of the touch sensor? The double-angle beams used for the bumper tend to wiggle a bit. Without those plates it would be possible for the bumper to get stuck on top of the touch sensor. The additional plates ensure that the double-angle beam remains "in its place" as it pivots.

A BETTER PROGRAM

We'll need to change the program slightly since the touch sensor is now pressed during normal operation and then released when the bumper hits something. This can be accommodated simply by changing the until statement:

```
until(BUMP==0);
```

Bumpbot is also a bit too predictable; it would be a little more interesting if it spun around a different amount each time it ran into something. The RCX has the ability to generate a random number between 0 and a number of our choosing. We can then use this number as the amount of time to spin. Unfortunately, that tends to result in Bumpbot's frequently spinning just extra a very small amount, and Bumpbot may take many tries to turn away from a wall. What we really want is for Bumpbot to always turn some minimum amount but also, to turn some additional random amount. We can do this in NQC simply by adding a constant value to the random value.

```
Wait(SPIN_MINIMUM + Random(SPIN_RANDOM));
```

Incorporating both of these changes, the new Bumpbot code looks like this:

```
// bumpbot2.nqc - improved sensor design

// sensors
#define BUMP SENSOR_1

// motors
#define LEFT  OUT_A
#define RIGHT OUT_C

// constants
#define REV_TIME      50
#define SPIN_MIN      70
#define SPIN_RANDOM   50

task main()
{
    // configure the sensor
    SetSensor(BUMP, SENSOR_TOUCH);
```

```
    // start going forward
    On(LEFT+RIGHT);

    // do this forever
    while(true)
    {
        // wait for bumper to hit something
        until(BUMP == 0);

        // back up
        Rev(LEFT+RIGHT);
        Wait(REV_TIME);

        // spin around
        Fwd(LEFT);
        Wait(SPIN_MIN + Random(SPIN_RANDOM));

        // resume
        Fwd(RIGHT);
    }
}
```

Similar changes can also be made to the RCX Code program.

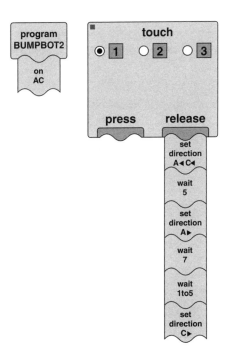

Figure 6–11:
Improved Bumpbot in RCX Code

CONCLUSION

With the ability to respond to its environment, Bumpbot is a bit more interesting than Tankbot. More sophisticated versions could be built with both front and rear bumpers, or perhaps a more complicated bumper that could discern between obstacles on the left or right side and respond accordingly. In any case, the basic ability to drive around while avoiding obstacles can be used as a stepping stone for more complicated vehicles.

Chapter 7

Bugbot

Bugbot once again extends the chassis introduced with Tankbot, this time by adding two antennalike feelers that give Bugbot its distinctive appearance. With its more sophisticated sensors and programming, Bugbot is much better at navigating around obstacles than the previous Bumpbots. The RCX's timer and counter capabilities are also utilized, and the concepts of *concurrent* and *synchronous* behavior are introduced.

CONSTRUCTION

In order to create Bugbot, you will first need to build Tankbot as shown in chapter 5. If you have just completed Bumpbot, you will need to undo those changes and restore Tankbot to its original form.

In the previous robots, motors and sensors were put in place, then wires were attached between the motors or sensors and the RCX itself. With Bugbot, however, the feeler assembly is built above the RCX input ports; therefore it is necessary to attach two wires (short ones will do) to inputs 1 and 3 ahead of time. Step 1 shows the addition of these wires and four *cross beams* to the original Tankbot. Cross beams are similar to regular beams except the hole is shaped like a cross instead of a circle. Because of this design, cross beams can be used to prevent an axle from rotating. They also hold an axle in place more firmly than regular beams do.

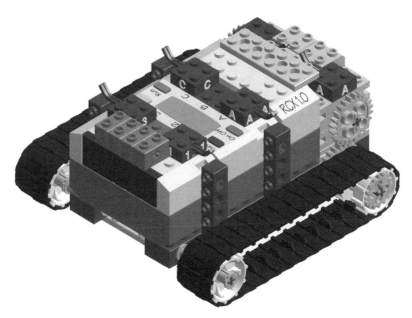

Figure 7-1:
Bugbot Step 1

Steps 2 and 3 build up the structure that will support the feelers. The pieces at the ends of the axle are called *1x3 crossblocks*. These pieces (and their shorter cousins) are quite versatile: one end has a cross hole (which prevents rotation), and the other end has a circular hole (allowing rotation). Bushings are used as spacers between the crossblocks and the cross beams. Without the spacers, it would be difficult to place both crossblocks the exact same distance from the cross beams, thus the feelers would not be symmetrical. A pair of touch sensors are also added; these will be activated by the feelers as Bugbot approaches an obstacle.

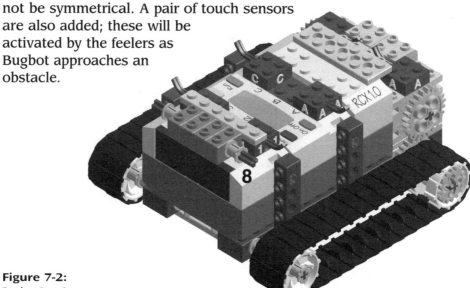

Figure 7-2:
Bugbot Step 2

Figure 7-3:
Bugbot Step 3

A #6 axle is passed through both touch sensors and secured in place by two bushings. These bushings are used later to hold two rubber belts in place. The feelers are built around #6 axles. Crossblocks are used to connect the axle to the peg that serves as a pivot point, and bushings are once again used as spacers. Rubber belts are used to hold the feeler against the touch sensor. The blue belts are a little too loose, but if they are wrapped a second time around the rear bushing the tension is just right.

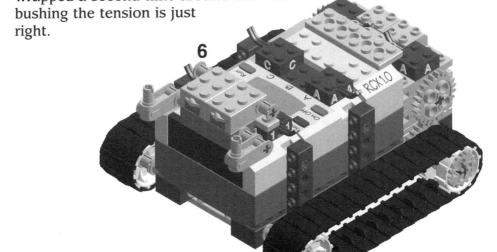

Figure 7-4:
Bugbot Step 4

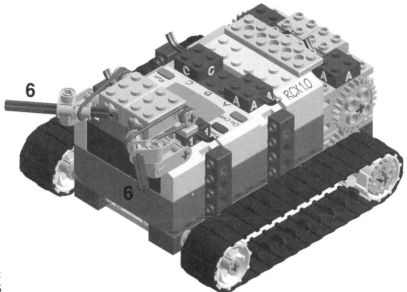

Figure 7-5:
Bugbot Step 5

Only a few final touches are needed. The touch sensors must be attached to the RCX using the wires from Step 1, and a piece with the unlikely name of *catch cross* is added to each of the feeler axles. This piece serves two purposes: it prevents the rubber belt from slipping off the feeler, and it provides better contact with the touch sensor than the axle alone. Flexible-ribbed tubing is added to extend the feelers (the longer magenta tubing works best). This tubing forms a sort of lever. The rubber belt is pulling on one side of the pivot, and an obstacle will push the feeler the other way. The distance from the pivot to the end of the tubing is much greater than the distance between the pivot and the rubber belt. Therefore, even a very small force applied to the tubing will counteract the rubber belt and cause the touch sensor to be released. The flexibility of the tubing, combined with the pivot design itself, gives the RCX plenty of time to react to obstacles before any serious impact is felt by Bugbot.

Figure 7-6:
Bugbot Step 6

RCX CODE BLOCKS

Before digging into the details of writing a program for Bugbot, a little discussion about RCX Code is in order. An RCX Code program is built up from individual *code blocks*. These blocks are grouped together into *stacks*.

> The RCX Code concept of stack should not be confused with the more traditional concept of a stack in computer science. In RCX Code, a stack is just a group of individual code blocks.

All programs start with a *program block* that contains the name of the program. The stack appearing under the program block will be executed whenever the program is run. The remaining code blocks are divided into four categories: *commands*, *sensor watchers*, *stack controllers*, and *my commands*.

Commands are colored green and represent a single action such as turning on a motor or playing a sound. In a way, commands do all of the real work of a program and the other types of blocks just serve to organize and coordinate the commands.

Sensor watchers are colored blue and are used to monitor a sensor (or other input source such as the timer or counter). The RCX continuously checks each sensor watcher, and if the appropriate condition is met, the stack below the watcher will be executed. Some sensor watchers have a single stack below them; others allow for two different conditions and can have two stacks. Sensor watchers always appear at the top of stacks; they cannot be placed underneath another block.

Stack controllers, colored red, are similar to sensor watchers. Many of the stack controllers check for conditions similar to those of a sensor watcher. These conditions can be used to choose one of two execution paths (**check & choose**), execute a group of blocks repeatedly (**repeat while**), or simply wait for a condition to be met (**wait until**). There are also stack controllers that repeat a group of commands without checking any sensors (**repeat** and **repeat forever**). When a group of commands is to be repeated, it is typically called a *loop*. Unlike sensor watchers, stack controllers appear in the middle (or at the bottom) of stacks, and not at the top.

The fourth category of blocks is *my commands*, and they are colored yellow. *My commands* allow you to add your own commands to an RCX Code program, then use those commands similarly to regular commands. In the original version of RCX Code, the definition of a *my command* starts with a "my commands" block that is very similar to a program block except that it is yellow and contains the text "my commands" followed by the name of the command (which you specify). A stack can be built below this block including regular commands and stack controllers. In version 1.5 of RCX Code, the definition of a *my command* has a starting and ending block (again in yellow), and commands are placed

My commands have a new look. **1.5** between these blocks. Apart from overall appearance, the big difference is that in version 1.5 a command's definition can be placed in the middle of some other stack.

The counterpart to a definition block is an invocation block, which looks very similar to a normal command block. The invocation block contains the name of the command and can be added to other stacks. You can even use the same *my command* several times by placing multiple invocation blocks within the other stacks. When using version 1.5, the definition itself can be placed within another stack, thus in some cases no invocation blocks will be

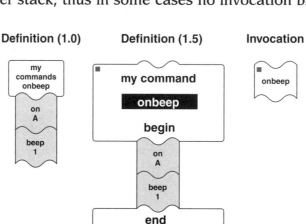

used. An example of a command definition (both in RCX Code 1.0 and 1.5) and an invocation appear below. For the remainder of the book 1.0 style illustrations will be used.

Figure 7-7:
"My Commands"

1.5 When using version 1.0, there are two significant restrictions for placing blocks in RCX Code programs. The first restriction is that a stack cannot contain more than one stack controller. For example, placing a **check & choose** controller in the same stack as a **repeat forever** controller is not allowed. The second restriction is that a *my command* cannot invoke another *my command* (or itself). Version 1.5 does not have such restrictions.

MULTIPLE WATCHERS

The idea behind Bugbot's programming is to have Bugbot drive forward until it hits something. If the obstacle is sensed by the right-hand feeler, then Bugbot should turn to the left while backing away slightly. If the obstacle is on the left, the Bugbot should back and turn to the right. The most obvious way to do this in RCX Code is to use a sensor watcher for each feeler. As you can see, backing and turning is accomplished by reversing one motor while stopping the other. We only back and turn until the feeler clears the obstacle.

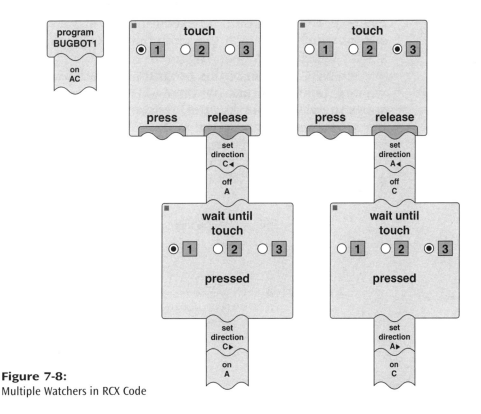

Figure 7-8:
Multiple Watchers in RCX Code

In order to completely understand the operation of the preceding program, it is important to remember that the RCX supports multi-tasking. A program may define multiple tasks, and any number of them may be active at a given time. When only a single task is active, its instructions are executed one after another. However, when multiple tasks are active, the RCX rotates through each of the active tasks, executing a single instruction from each task, then proceeding to the next one. Conceptually, this is similar to playing a board game with several people. Each person gets a turn and makes one move, then control passes to the next player. After everyone has had a turn, the first player gets to move again. The RCX adds the additional twist that tasks may become active or inactive as a program executes. Returning to our board game analogy, this is the equivalent of players leaving the game (tasks becoming inactive) or new players entering the game (tasks becoming active).

In RCX Code, the sensor watchers themselves are monitored in one task, while the stacks beneath the watchers are each executed in their own separate tasks (one task per stack). Thus, in the above example, the RCX could be executing the stack below the first sensor watcher while at the same time checking to see if the second sensor watcher has been triggered.

An NQC version of this program appears below. Note how each watcher is its own task (within a while loop), and the main task starts the watcher tasks after performing initialization (configuring the sensors and starting the motors). This is certainly more cumbersome than using if statements (as was done with Bumpbot), but it is a more accurate representation of the RCX Code program, and using separate tasks is relevant to our discussion.

```
// bugbot1.nqc - using multiple tasks

// sensors
#define LBUMP SENSOR_1
#define RBUMP SENSOR_3

// motors
#define LEFT OUT_A
#define RIGHT OUT_C
```

```
task main()
{
    // configure the sensor
    SetSensor(LBUMP,  SENSOR_TOUCH);
    SetSensor(RBUMP,  SENSOR_TOUCH);

    // start going forward
    On(LEFT+RIGHT);

    start watch_left;
    start watch_right;
}

task watch_left()
{
    while(true)
    {
        until(LBUMP == 0);
        Off(LEFT);
        Rev(RIGHT);

        until(LBUMP == 1);
        On(LEFT);
        Fwd(RIGHT);
    }
}

task watch_right()
{
    while(true)
    {
        until(RBUMP == 0);
        Off(RIGHT);
        Rev(LEFT);

        until(RBUMP == 1);
        On(RIGHT);
        Fwd(LEFT);
    }
}
```

Try running either of these programs on Bugbot and see how well it can maneuver. Overall, it does pretty well when it hits an obstacle with one feeler or the other, such as when approaching a wall at an angle. However, when heading straight into a wall, it has a tendency to stop dead in its tracks. Why does this happen? Nowhere in the program have we instructed Bugbot to turn off both of its motors.

The problem arises from the fact that the actions associated with the right and left sensors may run concurrently (at the same time). In the NQC program, the watch_left and watch_right tasks are both active. In the RCX Code program, the two sensor watchers are active at the same time. Now consider what happens if the two feelers come into contact with an obstacle at about the same time. Let's say the left feeler is detected first. The watch_left task (or the watcher on sensor 1 in RCX Code) will stop the left motor, reverse the direction of the right motor, and wait for the feeler to return to normal. Now what happens if the right feeler suddenly is noticed (because the watch_right task or second sensor watcher decided to check)? The right motor will be stopped, and the left motor will be reversed. Of course, the left motor is already stopped, so reversing it has no effect. The net result is that both motors are stopped and both watchers are waiting for sensors to be pressed, which, of course, won't happen since Bugbot isn't moving away from the obstacle. This is an example of the type of subtle problems that make it difficult to program and debug concurrent systems.

Perhaps you think the problem can be fixed by making the tasks (or watchers) turn the motor on in addition to reversing its direction before waiting for the sensor. In NQC, the fix would look something like this:

```
task watch_left()
{

    while(true)
    {
        until(LBUMP == 0);
        Off(LEFT);
        Rev(RIGHT);
        On(RIGHT);
```

```
                until(LBUMP == 1);
                On(LEFT);
                Fwd(RIGHT);
            }
        }

    task watch_right()
    {
        while(true)
        {
            until(RBUMP == 0);
            Off(RIGHT);
            Rev(LEFT);
            On(LEFT);

            until(RBUMP == 1);
            On(RIGHT);
            Fwd(LEFT);
        }
    }
```

Although this will prevent Bugbot from stopping completely, we will still have a potential problem when both feelers hit an obstacle. Basically, the watchers are fighting against one another. Each one is trying to correct its own situation but is making the other one's situation worse. In many cases, the watchers will fight back and forth, causing Bugbot to bounce back and forth until one of the watchers eventually "wins" and Bugbot returns to normal. In some situations this may be completely acceptable, but sometimes more dependable behavior is needed.

SYNCHRONOUS BEHAVIOR

An alternative to the multiple watcher problem is to ensure that only one feeler is being reacted to at a time and that we clear the condition on that feeler before checking the other one. This is called *synchronous* behavior. The idea is to have our program consist of a loop (sometimes called the *main loop*) which checks for the sensors being released, then responds to these events by executing the appropriate code.

In RCX Code, the main loop consists of a **repeat forever** in the program stack, and the sensors are checked using stack controllers (rather than sensor watchers). The start of such a program might look something like this (it only checks one sensor so far):

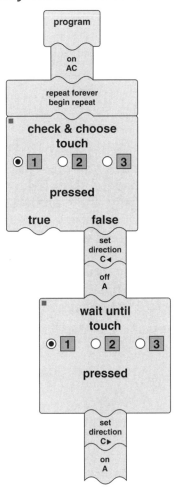

Figure 7-9:
Creating a Main Loop in RCX Code

Unfortunately, this code cannot be created in RCX Code 1.0. One restriction of RCX Code 1.0 is that a stack can contain at most one stack controller. The program stack above has three (**repeat forever**, **check & choose**, and **wait until**), and we haven't even handled the other sensor. We can begin to work around this problem by moving the **check & choose** (and the stack under it) into a *my command*, then invoking the *my command* under the **repeat forever**. Unfortunately the *my command* would then still have two stack controllers (**check & choose** and **wait until**). It is tempting to use the same " trick" again—that is, we could move the **wait until** into a new *my command* and invoke that second command from within the first. Unfortunately, another RCX Code 1.0 limitation gets in the way: *my commands* cannot be invoked from within other *my commands*. Faced with this limitation, we'll just have to adjust our program a little and turn for a fixed amount of time rather than waiting for the sensor to be pressed. The complete program is shown below:

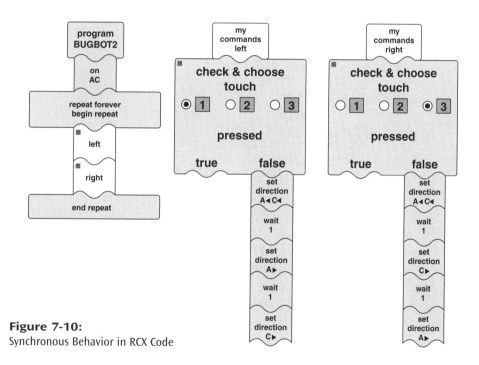

Figure 7-10:
Synchronous Behavior in RCX Code

NQC doesn't suffer from the same limitations as RCX Code 1.0—the various control structures such as if and while can be used as desired. For readability, however, we will still separate out the left and right sensor checking into functions called check_left and check_right. The revised program is shown below:

```
// bugbot2.nqc - synchronous behavior

// sensors
#define LBUMP  SENSOR_1
#define RBUMP  SENSOR_3

// motors
#define LEFT   OUT_A
#define RIGHT  OUT_C

void check_left()
{
    if (LBUMP == 0)
    {
        Off(LEFT);
        Rev(RIGHT);
        until(LBUMP == 1);
        On(LEFT);
        Fwd(RIGHT);
    }
}

void check_right()
{
    if (RBUMP == 0)
    {
        Off(RIGHT);
        Rev(LEFT);
        until(RBUMP == 1);
        On(RIGHT);
        Fwd(LEFT);
    }
}
```

```
task main()
{
    // configure the sensor
    SetSensor(LBUMP,  SENSOR_TOUCH);
    SetSensor(RBUMP,  SENSOR_TOUCH);

    On(LEFT+RIGHT);

    while(true)
    {
        check_left();
        check_right();
    }
}
```

STUCK IN A CORNER

Bugbot has a tendency to get trapped in a corner. The reason for this is that the right feeler will hit, Bugbot will turn left, then the left feeler hits, Bugbot turns right, and so on. Taking care of the corner problem requires two things: detecting that Bugbot is in a corner, and then taking corrective action. The corrective action is relatively easy: backing up and spinning around a bit should get Bugbot headed away from the corner.

Detecting the corner, however, is a bit more troublesome. The feelers don't know the difference between a corner and any other obstacle. All they know is if the obstacle is on the right, or the left, or possibly both. Corners can seem like any of these. Clearly we need some other way of deciding that we're in a corner.

When Bugbot encounters an obstacle, it usually can deal with it. There will be a little course correction to the right or left—sometimes even a few such corrections are needed—but eventually Bugbot heads away from the obstacle. A corner, however, causes Bugbot to thrash about: turning right, then left, then right again, never making much progress. Perhaps that is how we can tell if Bugbot is in a corner—we just need to determine when Bugbot is thrashing about. Instead of trying to detect the cause (being in a corner), we will attempt to detect the symptom (thrashing back and forth).

One way to do this is to keep track of how often Bugbot hits something. If we ever see a large number of hits within a fixed time period, then we'll assume that Bugbot is stuck and that is it time to panic. To make this determination, we need to introduce two new programming features: *variables* and *timers*.

The RCX contains 32 variables, which are a kind of memory. Each variable can hold an integer value in the range –32,768 to +32,767. The RCX also has functions that operate on the variables. For example, you can set a variable to a specific value, add a number to a variable, or check to see if a variable's value is equal to some other value.

RCX Code only allows the programmer to use a single variable—called the *counter*—and the operations allowed on the counter are restricted to the following:

- reset counter (set to 0)

- add to counter (increments the value by 1)

- check to see if the counter is within a specified range

NQC allows programmers to use up to 32 variables and provides access to a full set of arithmetic operations including addition, subtraction, multiplication, and division. The operands for these operations can be constants, other variables, or even dynamic values such as a sensor's reading or a random number.

We can use a variable to keep track of how many times Bugbot has hit an obstacle simply by incrementing the variable each time the right or left feeler is triggered. However, for our corner detection to work we also need a way to reset the variable to 0 every so often. This is where timers come in.

The RCX contains four timers that, like variables, are memory locations that hold a value which can be examined and checked. Unlike variables, timers change their value automatically; every tenth of a second each timer is incremented by 1. A program may reset a timer back to 0 at any time. After doing this, the timer will simply continue counting up from 0. This allows us to perform periodic tasks: if we need to do something every 3 seconds, then we create a task that waits for a timer's value to be 30 or more, reset the timer, perform the desired action, and go back to waiting for the timer. NQC allows access to all four timers, whereas RCX Code provides access to only a single timer.

Using variables and timers, our new program for Bugbot is shown on the following page. The program plays a short tone every time the timer task is triggered. This has no real function other than to let the user know that the timer task has been triggered.

Sometimes adding a few tones or beeps within a program can be very helpful while trying to debug it. In this case, the tone gives an indication of how often the counter is being reset.

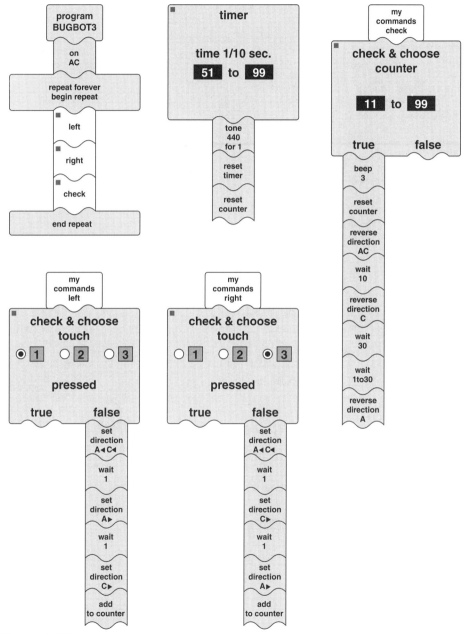

Figure 7-11:
Corner Detection in RCX Code

```nqc
// bugbot3.nqc - now with corner detection

// sensors
#define LBUMP SENSOR_1
#define RBUMP SENSOR_3

// motors
#define LEFT  OUT_A
#define RIGHT OUT_C

#define TIMER_LIMIT   50
#define COUNTER_LIMIT 10

int counter;

void check_left()
{
    if (LBUMP == 0)
    {
        Off(LEFT);
        Rev(RIGHT);
        until(LBUMP==1);
        On(LEFT);
        Fwd(RIGHT);
        counter += 1;
    }
}

void check_right()
{
    if (RBUMP == 0)
    {
        Off(RIGHT);
        Rev(LEFT);
        until(RBUMP == 1);
        On(RIGHT);
        Fwd(LEFT);
        counter += 1;
    }
}
```

```
void check_counter()
{
    if (counter > COUNTER_LIMIT)
    {
        // we're in a corner...
        PlaySound(SOUND_DOWN);
        counter = 0;

        // back up, spin around, and continue
        Rev(RIGHT+LEFT);
        Wait(100);
        Fwd(RIGHT);
        Wait(300 + Random(300));
        Fwd(LEFT);
    }
}

task main()
{
    counter = 0;

    // configure the sensor
    SetSensor(LBUMP, SENSOR_TOUCH);
    SetSensor(RBUMP, SENSOR_TOUCH);

    On(LEFT+RIGHT);
    start watch_timer;

    while(true)
    {
        check_left();
        check_right();
        check_counter();
    }
}

task watch_timer()
{
    while(true)
    {
```

```
            until(Timer(0) > TIMER_LIMIT);
            PlayTone(440, 10);
            ClearTimer(0);
            counter = 0;
        }
    }
```

Note how the counter is checked within the main loop, but the timer is watched in its own task. This is because after the counter check, the program may need to control the motors. If this check were done in its own task, we would encounter problems similar to the very first Bugbot program. For example, consider what would happen when Bugbot was stuck in a corner. The main loop would be checking the right and left feelers, and it would be responding by turning Bugbot appropriately and incrementing the counter. At some point, the counter would exceed the limit set for corner detection. The code responsible for checking the counter would want to make Bugbot back away from the corner. However, if this code were in its own task, the main loop would continue to check the feelers while this was happening. Depending on the circumstances, one of the feelers may have hit an obstacle before the counter checking task had reversed the motors. In this case, the feeler checking code would be fighting against the counter checking code. The results of two tasks fighting for control, as shown in our first Bugbot example, are unreliable at best.

The timer, however, does not directly interact with the motors, so there are no adverse side effects to letting it remain in its own task. In fact, keeping it a separate task actually makes for a cleaner program.

TUNING THE CORNER DETECTOR

Our corner detection algorithm relies on two parameters: how many obstacle hits constitute a "corner" and how often this counter should be reset. The above program uses a limit of 10 hits within a 5-second period. Depending upon the type of obstacles Bugbot must negotiate, some trial and error may be required to determine the right values. The remaining paragraphs will refer to these values by their NQC names: COUNTER_LIMIT and TIMER_LIMIT. If you are using RCX Code, then COUNTER_LIMIT is the lower value used in the counter **check & choose** block, and TIMER_LIMIT is the lower value used in the timer watcher.

To find a set of values, first pick an amount of time that you are willing to let Bugbot thrash about in a corner before panicking and use that to set TIMER_LIMIT (actually Bugbot may thrash in the corner for twice that much time in the worst case). Now set COUNTER_LIMIT to something large (99 is a nice number) and send Bugbot running into a corner. Listen for the tones that indicate the timer triggering and count the number of collisions that occur between those tones. For best results you should repeat this several times, then set COUNTER_LIMIT to one less than the smallest count you made.

Now try running Bugbot around the room and see how it behaves. Try sending it into a corner and listen for the panic sound. If you don't hear the panic sound, then COUNTER_LIMIT is too high; try using a slightly smaller value. In the other extreme, if Bugbot panics even when negotiating normal obstacles such as a wall, then TIMER_LIMIT needs to be increased and a new value for COUNTER_LIMIT should be found.

If Bugbot is still panicking too frequently, a small delay can be inserted into the turning code so that Bugbot turns a little more after each collision, and presumably will negotiate obstacles with fewer collisions. For example, the check_left function for the NQC program could be modified as follows (with a similar modification to check_right):

```
#define TURN_DELAY 10

void check_left()
{
    if (LBUMP == 0)
    {
        Off(LEFT);
        Rev(RIGHT);
        until(LBUMP == 1);
        Wait(TURN_DELAY); // turn a little extra
        On(LEFT);
        Fwd(RIGHT);
        counter += 1;
    }
}
```

CONCLUSION

The mechanical design of Bugbot was simple enough, but it led us to some rather complicated programming issues. Specifically, using a concurrent behavior led to some subtle problems that could be avoided with a synchronous behavior. In addition, the timer and counter facilities of the RCX were introduced to solve the problem of getting stuck in a corner.

Some of the limitations of RCX Code were also encountered. In some cases, these limitations can be worked around with a little bit of ingenuity, but in many cases NQC provides a more elegant way to tap the full potential of the RCX.

Chapter 8

Linebot

Linebot is a robot that follows a path drawn on the floor. This simple behavior can be used as a building block in more complex robots. Linebot introduces the light sensor and also explains the concept of *feedback*, and how it can be used in programming robots.

CONSTRUCTION

Following a line requires more precise movement than our previous robots, so the Tankbot chassis will need some minor modification. Instead of driving the treads using a 1:1 gear ratio, Linebot will use a 3:2 ratio, which results in a slightly slower speed. The only other modification is the mounting of a light sensor at the front of Linebot.

Construction starts with Tankbot stripped down to its bottom frame (Tankbot Step 3 minus the 2x8 plate).

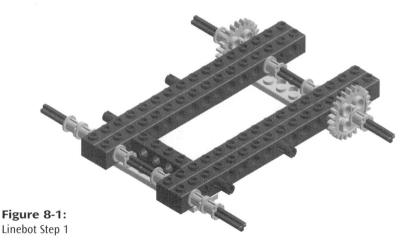

Figure 8-1:
Linebot Step 1

Tankbot used 24-tooth gears on the rear axles and the motors (1:1 gear ratio). For Linebot, we want to add some gear reduction, so 16-tooth gears will be used on the motors while 24-tooth gears remain on the axles (3:2 gear reduction). Because the 16-tooth gear has a smaller radius than the 24-tooth gear, the motors must be placed slightly lower than before in order for the gears to mesh properly. The space between the frame and the motors is now only one plate height instead of two. Steps 2 and 3 illustrate this revised motor arrangement. Note that since the motors are one plate lower than before, an extra 1x4 plate (for a total of two) must be added on top of the motors to make them flush with the RCX.

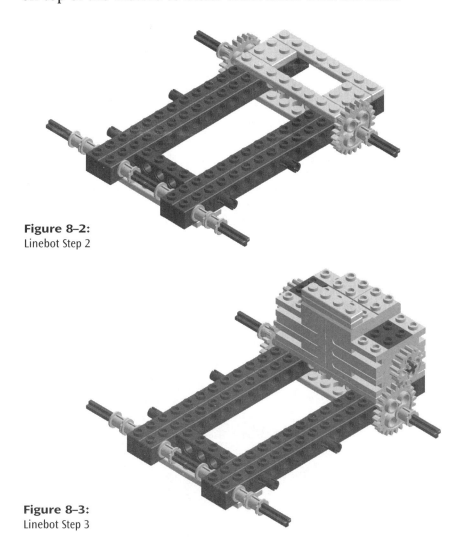

Figure 8–2:
Linebot Step 2

Figure 8–3:
Linebot Step 3

The light sensor itself is positioned just barely above the floor. The light sensor responds to *ambient light* (the overall light in the room) as well as to *reflected light* from its own LED. By placing the light sensor close to the floor, the amount of ambient light is minimized, thus increasing the overall reliability of measuring reflected light. When attempting to detect reflected light, it is a good design practice to get the sensor as close as possible to the object to be seen.

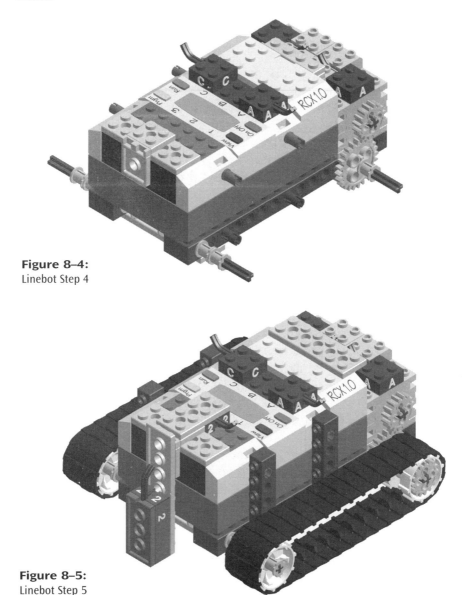

Figure 8–4:
Linebot Step 4

Figure 8–5:
Linebot Step 5

In contrast to some of the previous designs (such as Bumpbot), Linebot is somewhat fragile (especially the mounting of the light sensor). Even the slightest impact will snap off the light sensor. A more robust mounting, perhaps utilizing cross beams and some vertical bracing, could be designed, but since we don't expect Linebot to bump into things during normal operation, we'll stick with this simple (yet fragile) design. A note of caution: don't try to use Linebot on a carpeted floor. The rough surface of the carpet will typically knock the light sensor off of Linebot.

SEEING THE LINE

The goal of Linebot's program is to allow it to follow the path of a black line drawn on a white surface. The test mat included with the Robotics Invention System can be used, or you can create your own test course. The line should be a solid dark color and at least ½″ wide. Although such a course can be drawn with markers, I prefer to cut out strips of construction paper and tape or glue them into place on a poster board. This provides nice even colors and crisp edges.

It is tempting to have the robot follow a line by attempting to keep the light sensor centered directly over the line. Since the light sensor can easily differentiate between black and white, the robot will be able to tell whether it is still following the line or has wandered astray. Unfortunately, just knowing that the robot has gone astray is not enough. It must also know whether to correct by turning to the right or to the left. Since both sides of the line look the same (white), what can the RCX do?

The answer has to do with the way that the light sensor sees things. The sensor measures the total amount of light falling on it. In the case of Linebot, most of this light is light from the built-in LED that is reflected back up from the surface. White paper reflects much of the light, while black absorbs most of it. But the sensor sees more than just a single point directly below it; light bouncing back anywhere in a small region will cause the sensor to respond. When this entire field of vision is over the white paper, the sensor reads a very high value. When the entire field is over the black line, a very low value is read. When the field is only partly over the line, however, an intermediate value is read.

Knowing this, we can change our strategy slightly. Instead of trying to keep the sensor centered over the line, we will try to keep it centered on the edge between the blank surface and the line.

When correctly centered we should read an intermediate value. We will detect a high value if the robot wanders too far to the left and a low value if it wanders too far to the right. Our program can then correct the situation by turning a little to the right or left to remedy the situation. This is an example of a *feedback system,* and it is an important enough concept to warrant some extra discussion.

FEEDBACK

Consider a household thermostat that has a single temperature setting. Let's say the thermostat is set for 70 degrees and it is winter time, so a furnace is being used to heat the house. When the temperature falls below 70 degrees, the thermostat starts the furnace, which slowly begins to heat the house. The furnace remains on until the temperature rises above 70 degrees, at which point the thermostat turns it off. The condition used to control the furnace (temperature) is the same thing that is actually affected by the furnace. This is called feedback, and it is a very powerful technique for programming robots. The operation of this thermostat is illustrated below:

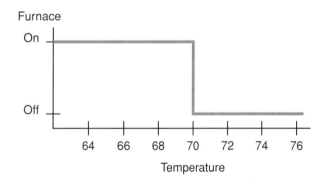

Figure 8-6:
Operation of Thermostat

When using feedback, it is often important to keep the system from reacting too quickly. Returning to our thermostat example, consider what would happen if the thermostat turned on the furnace as soon as the temperature was 69.9999 degrees, then turned it off the moment it hit 70.0000. The furnace would be perpetually turning on and off rapidly, with the slightest air current causing it to change state.

To make matters worse, the temperature of the houe would still oscillate a few degrees above or below 70 degrees. This is because houses rarely heat in a uniform manner and a good deal of air circulation is usually necessary to average out the air temperature. In addition, it takes a certain amount of time for the effects of the furnace to be detected by the thermostat.

A better solution is to make the threshold for turning on the furnace a few degrees cooler than the threshold for turning it off. For example, we could turn on the furnace when it is 68 degrees, and turn it off when it is 72 degrees. For any value in between (e.g., 70 degrees), the furnace would remain in its current state. This technique is known as *hysteresis* and is illustrated below:

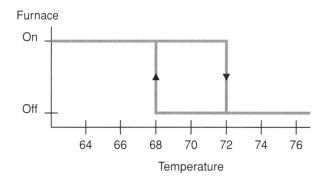

Figure 8–7:
Adding Hysteresis to a Thermostat

This graph is a bit unusual in that there are two possible states for the furnace (on and off), when the temperature is between 68 and 72 degrees. This is because the decision for turning the furnace on or off depends both on the current temperature and, whether or not the furnace is already on. Consider what happens when the temperature is 64 degrees. The furnace will be running and will begin to heat up the house. Even when the temperature is 70 degrees the furnace will remain on. Only when the temperature rises above 72 degrees will the furnace turn off. Presumably the house will then slowly begin to cool, but rather than turning on the furnace right away, the thermostat waits until the temperature drops to 68 degrees.

Most thermostats use some form of hysteresis, although it usually isn't as extreme as in our example.

FOLLOWING THE LINE

As discussed above, when the light sensor is completely over the black surface, it will read a low value. When the light source is over the white surface, it will register a significantly higher value. While it is over the edge between black and white, it will register a value somewhere in the middle. Since we are going to be using a feedback algorithm with hysteresis, we will need two threshold values: one for turning left, the other for turning right.

As long as the sensor reads between these two thresholds, the robot can continue moving straight ahead. If the sensor reads too low, then the robot is too far to the right and needs to turn left. This can be accomplished simply by turning off the left motor. The case where the sensor reads too high is similar, but the right motor is turned off instead.

Specific threshold values can be determined by using the view function of the RCX to monitor the light sensor's value while manually positioning it over the white surface or the black line. To view the sensor's value, the RCX must first be told that sensor 2 is a light sensor. One way to do this is to download and run a program that uses sensor 2 as a light sensor—any of the programs in this chapter will work fine. The program doesn't have to run for very long; just starting it will configure the RCX properly. Pressing the **View** button twice will then switch the RCX from the default display mode to viewing sensor 2.

The threshold for turning left should be a little larger than the reading over the black line. The threshold for turning right should be a little smaller than the reading over the white surface. For example, if the sensor reads a value of 56 over the white surface and 39 over the black line, then thresholds of 53 and 42 could be tried. When thresholds are close together, Linebot is pickier about what it considers "over the line," and this can result in a lot of adjustments as it attempts to follow the line. On the other hand, thresholds that are too far apart run the risk of letting Linebot wander too far astray. The light sensor readings are susceptible to many factors including the ambient lighting in the room, the condition of the RCX's batteries, and the exact colors and surfaces used. Because of this, proper calibration of the thresholds will often require a bit of trial and error. If you observe that Linebot is constantly adjusting back and forth, even over a straight line, then the thresholds are probably too close together. If Linebot tends to wander completely off the line (perhaps even spinning in circles), the thresholds are probably too far apart.

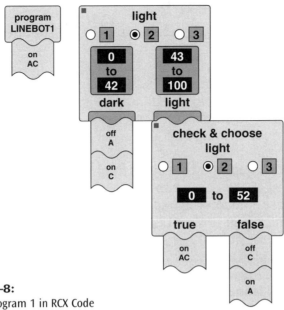

Figure 8–8:
Linebot Program 1 in RCX Code

```
//linebot1.nqc
//turn by stopping one tread

// sensor and motors
#define EYE     SENSOR_2
#define LEFT    OUT_A
#define RIGHT   OUT_C

//thresholds
#define LEFT_THRESHOLD      42
#define RIGHT_THRESHOLD     53

task main()
{
    SetSensor(EYE,  SENSOR_LIGHT);
    On(LEFT+RIGHT);

    while(true)
    {
        if (EYE <= LEFT_THRESHOLD)
        {
```

```
            Off(LEFT);
            On(RIGHT);
        }
        else if (EYE >= RIGHT_THRESHOLD)
        {
            Off(RIGHT);
            On(LEFT);
        }
        else
        {
            On(LEFT+RIGHT);
        }
    }
}
```

Depending upon the condition of Linebot's batteries and the surface it is driving on, turning may become erratic. This is because turning requires the treads to slide sideways across the ground. It turns out that turning is a little smoother if both treads are running. One way to do this is to have one tread running forward while the other one is in reverse. If you are using NQC, try modifying the `linebot1.nqc` program to do this (hint: use `Rev` and `Fwd` instead of `Off` and `On` within the `while` loop).

In many cases, this simple change results in Linebot rapidly spinning back and forth without making any forward progress. This is indicative of a feedback system that reacts too quickly and/or overreacts. If there is enough contrast between the black line and the white background, you may be able to solve this problem by moving the thresholds further apart.

A second option is to make sure that each time a course correction is required, the correction is completed and at least some forward progress is made before checking the sensor again. In effect, we are slowing down the feedback loop and making it a little less responsive (which is another form of hysteresis). The NQC version of this approach is shown below. Note that Linebot will now be turning faster and reacting slower than in the previous example, thus it may be necessary to use a wider line. If the line is too thin, Linebot may overshoot it and get lost.

```
//linebot2.nqc
//turn by reversing one tread

// sensor and motors
#define EYE        SENSOR_2
#define LEFT       OUT_A
#define RIGHT      OUT_C

// thresholds
#define LEFT_THRESHOLD   42
#define RIGHT_THRESHOLD 53

#define STRAIGHT_TIME 10

task main()
{
    SetSensor(EYE, SENSOR_LIGHT);
    On(LEFT+RIGHT);

    while(true)
    {
        if (EYE <= LEFT_THRESHOLD)
        {
            Rev(LEFT);
            until(EYE > LEFT_THRESHOLD);
            Fwd(LEFT);
            Wait(STRAIGHT_TIME);
        }
        else if (EYE >= RIGHT_THRESHOLD)
        {
            Rev(RIGHT);
            until(EYE < RIGHT_THRESHOLD);
            Fwd(RIGHT);
            Wait(STRAIGHT_TIME);
        }
    }
}
```

Note how the turns have their own until conditions so that they turn far enough to be back on course. Afterward, Linebot proceeds straight ahead for a fixed amount of time (determined by the constant STRAIGHT_TIME) before checking the sensor again. A high value for STRAIGHT_TIME will work better for relatively straight paths, while lower values work better for sharp curves. Choosing the appropriate value for such a constant is part of the art of robot design.

It is difficult to implement the same algorithm in RCX Code 1.0 (however, it can be implemented in RCX Code 1.5). Due to the limitations on nesting stack controllers, we cannot have the "right turn" and "left turn" run in the same stack. Instead, they must appear as separate stacks underneath a sensor watcher. The program for this is shown below:

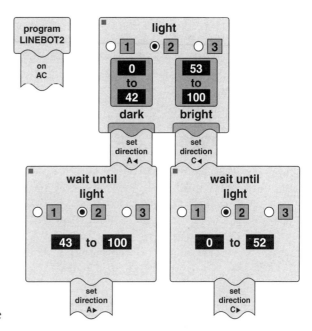

Figure 8–9:
Linebot Program 2 in RCX Code

Unlike the NQC program, there is no **wait** command after each turn is made. This is because such a command would typically have no effect in the above program. Keep in mind that stacks under a sensor watcher each run in their own tasks. A separate task is used to implement the watcher itself—let's call it task "A." Consider a typical scenario where Linebot has drifted too far into the line, thus the sensor reading is below the threshold. Task "A" would notice that the current light sensor value is in the range 0 to 41 and would thus start the task associated with the left half of the sensor watcher, which we'll call task "B." Motor A would start running backward, and task "B" would wait for a brighter light sensor reading. At some point, Linebot will have turned far enough to read a lighter value and motor A will start running forward again. If we placed a **wait** command at the bottom of the stack, then task "B" would have to wait a certain amount of time before continuing. However, task "A" would still be free to do whatever it wanted during this time. In particular, if the light sensor reading got too bright, it would then start task "C" to run the actions associated with the right half of the sensor watcher. The result is that Linebot will tend to turn back and forth quite rapidly, and there's not a lot we can do about it.

With the right amount of contrast between black and white and appropriate thresholds, the above program can work reasonably well. However, in some situations the RCX Code program responds too slowly in the **wait until** blocks and tends to overcompensate its steering. When this happens, Linebot will spend most of its time zigzagging back and forth rather than making forward progress. In these cases the speed of the motors can be reduced (using a **set power** command), or the original program may be used instead.

IMPROVED LINE FOLLOWING

Our previous tracking algorithm tried to keep the robot positioned over the edge between a white surface and a black line. It is also possible to design an algorithm to follow the line itself, although this requires using multiple variables and cannot be done within the confines of RCX Code.

This added complexity, however, is not without its rewards. If we make the line itself gray rather than black, and design an algorithm to follow the gray line, then we can also place black *stoppers* along the line and instruct Linebot to take some action whenever it encounters one of these stoppers. The theory behind this algorithm is to drive forward until the light sensor exceeds some preset threshold indicating that the robot has wandered off the line. The robot will then start searching alternately right and left looking for the line. Once it finds the line, it will continue moving forward.

```
// linebot3.nqc
// improved tracking algorithm with stoppers

// sensor and motors
#define EYE      SENSOR_2
#define LEFT     OUT_A
#define RIGHT    OUT_C
```

```
// light sensor thresholds
#define LINE_THRESHOLD        51
#define STOPPER_THRESHOLD     42

// other constants
#define TURN_SPEED 3
#define INITIAL_TIME 4

int direction, time, eye, ok_to_stop;

void follow_line()
{
    // initialize the variables
    direction = 1;
    time = INITIAL_TIME;
    ok_to_stop = 0;

    // start driving
    OnFwd(LEFT+RIGHT);

    while(true)
    {
        // read the sensor value
        eye = EYE;

        if (eye < LINE_THRESHOLD)
        {
            // we are either over the line or a stopper
            if (eye > STOPPER_THRESHOLD)
                ok_to_stop = 1;
            else if (ok_to_stop == 1)
            {
                // found a stopper
                Off(RIGHT+LEFT);
                return;
            }
        }
        else
        {
            // need to find the line again
            ClearTimer(0);
            if (direction == 1)
```

```
        {
            SetPower(RIGHT+LEFT, TURN_SPEED);
            Rev(LEFT);
        }
        else
        {
            SetPower(RIGHT+LEFT, TURN_SPEED);
            Rev(RIGHT);
        }

        while(true)
        {
            // have we found the line?
            if (EYE < LINE_THRESHOLD)
            {
                time = INITIAL_TIME;
                break;
            }

            if (Timer(0) > time)
            {
                // try the other direction
                direction *= -1;
                time *= 2;
                break;
            }
        }

        SetPower(RIGHT+LEFT, OUT_FULL);
        Fwd(RIGHT+LEFT);
    }
  }
}
```

As usual, we start with the definitions of constants and variables for the program. The line tracking algorithm itself is contained within the function follow_line, which will track a line until it encounters a stopper, at which point it will stop the motors and return. The basic operation of follow_line is to initialize the variables, start moving forward, then enter the main loop. This loop reads the light sensor, then decides if the sensor is over the line or not. If it is over the line, then it must further decide if a stopper has been encountered. If it is not over the line, then it must search for the line.

Although it is easy to detect a stopper (check to see if the light sensor is below a given threshold), this feature introduces a subtle complication. Consider what would happen after the first stopper is encountered. The `follow_line` function would stop the motors and return to its caller. Now, suppose that the caller wants to move on to the next stop. It would call `follow_line` again, but since the light sensor is still over the stopper, the function would immediately return. We need some way of knowing if the stopper is the same one we started on or a new one. The algorithm uses the `ok_to_stop` variable to capture this information. Initially `ok_to_stop` is set to 0 (indicating that it is not okay to stop). If the light sensor ever reads a value between `LINE_THRESHOLD` and `STOPPER_THRESHOLD`, then Linebot must be over a portion of the line that is not a stopper, and it sets `ok_to_stop` to 1. The algorithm only pays attention to stoppers when `ok_to_stop` is 1, which in effect disables stopper detection until some part of the line other than a stopper has been detected.

Each search for the line results in a turn (either left or right), which continues either until the line is detected or a certain amount of time elapses. Each successive pass results in a longer maximum turn time in the opposite direction. The variables `direction` and `time` hold the present direction and time limit for searching. The search begins with resetting the timer and instructing the motors to turn. Turns are made at partial power with one tread reversed. After the turn has been started, a loop keeps checking to see if either the line has been detected or the timeout has expired. If the line is detected, then the `time` is reset to its initial value so that the next search will once again start with small turns. If the timeout is exceeded, then `direction` is reversed and `time` is doubled, so that the next pass through the main loop will result in a larger turn in the opposite direction.

Reversing `direction` and doubling `time` uses an operation that is somewhat peculiar:

```
direction *= -1;
time *= 2;
```

The `*=` operator is a special operator that multiplies a variable by a value and places the result back into the variable. There are corresponding operators for the other basic arithmetic operations as well (`+=`, `-=`, and `/=`).

Those statements could have also been written this way:

```
direction = direction * -1;
time = time * 2;
```

Many C programmers (myself included) prefer using the special operators such as *= because they are more compact and easier to maintain (if you want to change the variable being modified, you only need to change it in one place, instead of in both places).

Using the algorithm is quite simple; all that is needed is to configure the light sensor and then call follow_line as desired. The following code will cause Linebot to follow the line, playing a sound, and pausing for one second whenever a stopper is encountered.

```
task main()
{
    SetSensor(EYE, SENSOR_LIGHT);

    while(true)
    {
        follow_line();
        PlaySound(SOUND_FAST_UP);
        Wait(100);
    }
}
```

As with the previous Linebot programs, creating a suitable test course and setting the proper thresholds may require some experimentation. The test course should be designed to have excellent contrast between the blank surface, the line to be followed, and the stoppers. It is also important to make the turns gradual and smooth. An extremely sharp turn may cause Linebot to lose track of the line. I constructed my test course on white poster board and used green construction paper for the line and black paper for the stoppers. The lines were 3/4" thick, and the sharpest turn used a 6" radius. The stoppers were 3/4" x 3/4". If you place the stoppers on or near a curve, then they may need to be a little longer than 3/4".

As for thresholds, the LINE_THRESHOLD should be midway between the reading for the background and the reading for the line. The STOPPER_THRESHOLD should be midway between the line reading and the stopper reading. For example, if the background reading is 56, the line reading is 46, and the stopper reading is 39, then thresholds of 51 and 42 can be used for LINE_THRESHOLD and STOPPER_THRESHOLD, respectively. If Linebot spends most of its time turning back and forth looking for the line, then the LINE_THRESHOLD may be too low. On the other hand, if it is perfectly content to wander away from the line without even trying to correct its course, then LINE_THRESHOLD is probably too high. For the STOPPER_THRESHOLD, if Linebot always misses the stopper, then the value is too low, otherwise if Linebot occasionally stops when no stopper is present, then the value is too high.

CONCLUSION

Several different algorithms for following a line were presented, and the concepts of feedback and hysteresis were introduced. The behavior of following a line is a good building block for more complicated robots. Consider a robotic garbage truck that follows a path on the floor, emptying any garbage cans that it encounters along the way. Another variation would be a robotic mailman that follows its route, dropping off mail at each stop. In the next chapter, a robotic dump truck, Dumpbot, will be built using Linebot as a starting point.

Chapter 9

Dumpbot

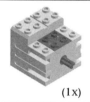

Dumpbot is a robotic dump truck based on Linebot. Like its predecessor, it can follow a line to drive from place to place. In addition, it has a motorized bin that can be used to transport cargo and dump it out when desired. Although Dumpbot can be programmed using RCX Code, a more advanced algorithm can be employed using NQC.

Extra Pieces Required:

1 Motor

See Appendix B for information on obtaining extra pieces.

(1x)

CONSTRUCTION

Because Dumpbot is based on Linebot, the first stage of construction is to add the appropriate support and connecting wire to Linebot so that the cargo bin may be attached later on.

Figure 9-1:
Dumpbot Step 1

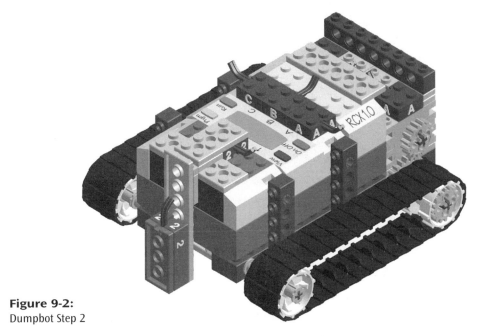

Figure 9-2:
Dumpbot Step 2

Steps 3 and 4 show the construction of the dumping mechanism. The 1×4 beams will later be attached to the cargo bin itself, and the #10 axle will be powered by a motor (after suitable gear reduction). The oval-shaped pieces added in Step 4 are called *1x3 liftarms*, and they are remarkably versatile. In this case they are being used just as their name implies: as an arm for lifting something. In later chapters, liftarms will be employed for bracing and other structural purposes as well.

Figure 9-3:
Dumpbot Step 3

Figure 9-4:
Dumpbot Step 4

The frame for the dumping mechanism is constructed in Steps 5 through 8. The two 1×2 plates near the back of the frame will serve as a stopper to limit the cargo bin's movement later on.

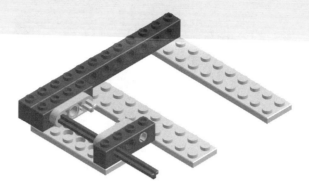

Figure 9-5:
Dumpbot Step 5

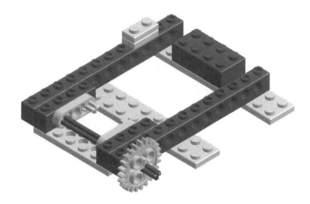

Figure 9-6:
Dumpbot Step 6

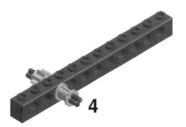

Figure 9-7:
Dumpbot Step 7

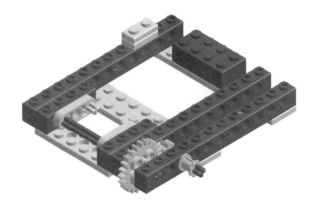

Figure 9-8:
Dumpbot Step 8

The 1×4 plates added in Step 9 are used to support the motor. As usual, we need to accommodate the unusual shape of the motor's bottom by providing support along only the edges of the motor.

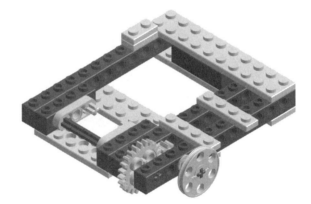

Figure 9-9:
Dumpbot Step 9

The cargo bin itself is built in Steps 10 through 12. The friction peg shown in Step 10 will hit the 1×2 stopper plates added back in Step 6. A 1×4 plate serves as a lip for the cargo bin so that cargo won't roll out until the bin is tipped. Depending on the exact cargo, a larger lip could be built (perhaps two 1×4 plates, or even a beam).

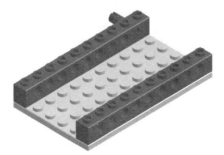

Figure 9-10:
Dumpbot Step 10

Figure 9-11:
Dumpbot Step 11

Figure 9-12:
Dumpbot Step 12

In Step 13, the cargo bin is attached to the 1×4 beams of the dumping mechanism. The motor is also added, with a blue belt to connect the pulleys. The pulleys provide approximately a 7:2 reduction, which, when combined with the 3:1 gear reduction, yields a total reduction of 21:2. This is suitable to lift and dump a modest amount of cargo quickly. The belt drive also allows the motor to slip when the cargo bin's motion is limited by the stopper.

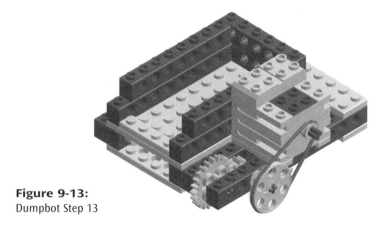

Figure 9-13:
Dumpbot Step 13

The dumping mechanism is attached to the back of Linebot in Step 14. Some minor adjustment of the tread motor wires (outputs A and C) may be required in order to get the dumping mechanism firmly in place. The dumping motor is connected to output B. The best route for this wire is to run it straight backward (over the top of the 1×8 beam), then wrap it up and around to the motor.

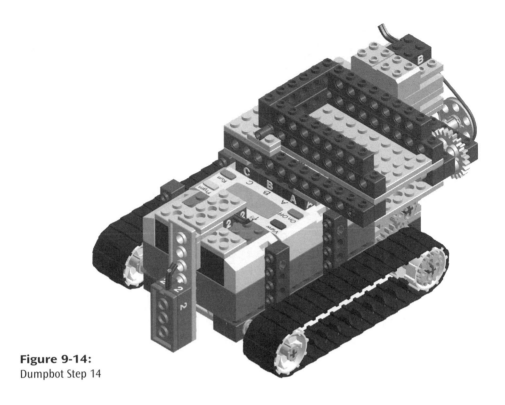

Figure 9-14:
Dumpbot Step 14

A SIMPLE PROGRAM

Since Dumpbot's construction is based on Linebot, it makes sense that its programs are also based on those from Linebot. Our first program is an easy one: follow a line for 5 seconds, then dump the cargo bin. To dump the cargo bin, the motor is run forward for 0.7 second to tilt the bin, then backward for 1.0 second to restore the bin to a horizontal position. The reason that the motor is run longer in reverse than forward is that there is always some variability in how fast the bin tips. If the forward and reverse times were exactly equal, there would always be the chance that the bin wouldn't get fully restored to a horizontal position. The stopper (the friction peg that hits the 1×2 plates) will prevent the bin from moving too far, and the rubber belt will allow slippage once this happens.

The following NQC code defines the appropriate constants and a function to dump the cargo bin:

```
// dumper motor and constants
#define DUMPER  OUT_B
```

```
#define DUMP_TIME      70
#define UNDUMP_TIME   100

void dump()
{
    OnFwd(DUMPER);
    Wait(DUMP_TIME);
    Rev(DUMPER);
    Wait(UNDUMP_TIME);
    Off(DUMPER);
}
```

The original line-following algorithm from Linebot is close to what we need, with one exception: we need to be able to stop it after 5 seconds. This can be accomplished by placing the follow routine in its own task. The main task can then start the follow task, wait 5 seconds, stop the follow task, then dump the cargo. The complete program appears below:

```
// dumpbot1.nqc
//  follow a line for 5 seconds, then dump cargo

// linebot sensor, motors, and constants
#define EYE        SENSOR_2
#define LEFT       OUT_A
#define RIGHT      OUT_C
#define LEFT_THRESHOLD      42
#define RIGHT_THRESHOLD     53

// dumper motor and constants
#define DUMPER       OUT_B
#define DUMP_TIME    70
#define UNDUMP_TIME  100

#define FOLLOW_TIME   500 // 5 seconds of driving before dumping

void dump()
{
    OnFwd(DUMPER);
    Wait(DUMP_TIME);
    Rev(DUMPER);
    Wait(UNDUMP_TIME);
```

```
        Off(DUMPER);
}

task main()
{
    SetSensor(EYE, SENSOR_LIGHT);

    start follow;
    Wait(FOLLOW_TIME);
    stop follow;
    Off(LEFT+RIGHT);
    dump();
}

task follow()
{
    On(LEFT+RIGHT);

    while(true)
    {
        if (EYE <=  LEFT_THRESHOLD)
        {
            Off(LEFT);
            On(RIGHT);
        }
        else if (EYE >=  RIGHT_THRESHOLD)
        {
            Off(RIGHT);
            On(LEFT);
        }
        else
        {
            On(LEFT+RIGHT);
        }
    }
}
```

A similar program can be created in RCX Code. Once again we start with the original Linebot program, then add the capability to dump the cargo bin and halt the line-following code. RCX Code does not allow direct control of tasks, so we need another way to stop following the line. Simply turning motors A and C off would work most of the time. The problem is that if the light sensor watcher is triggered at just the right time, it might turn one or more of the motors back on. Consider the following scenario: Dumpbot has just wandered a little too far over the line, which triggers the stack under the "dark" portion of the sensor watcher. This stack starts to run and turns off motor A. Before it can get to the next step, however, the main program notices that 5 seconds have elapsed and manually turns off motors A and C. Now the light sensor task finishes running and turns motor C back on. This is another example of the subtle difficulties of concurrent systems, as discussed in chapter 7.

One solution is to keep turning motors A and C off just in case one of the other stacks is still running. The counter is used to decide if the motors need to be turned off or not. The main program increments the counter, which triggers the counter watcher stack to keep turning the motors off. The complete program is shown below:

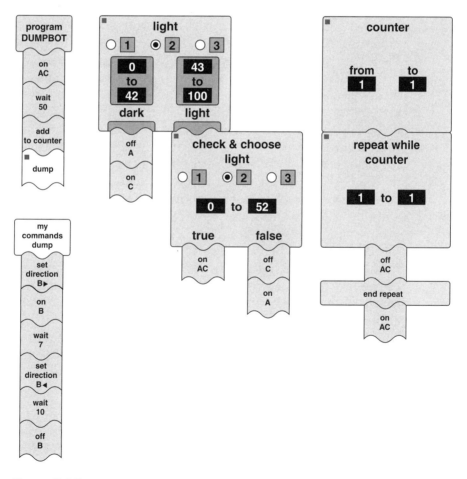

Figure 9-15:
Dumpbot Program in RCX Code

A few simplifications could be made to this program. For example, the first **set direction** in the **dump** command isn't really necessary since the motors always default to the forward direction when a program starts. However, including the **set direction** command makes the **dump** command more versatile since it can be called multiple times within a program. Similarly, since the counter is never reset within the program, the counter watcher could be simplified by changing the **repeat while** to a **repeat forever** and discarding the **on** command. However, a more complex program may choose to continue following the line after dumping the cargo, which can be accomplished by resetting the counter.

Whichever program you use (NQC or RCX Code), you'll need to create a line for Dumpbot to follow. You can use the test mat included with the Robotics Invention System or create your own. Whatever you used to test the first version of Linebot should work for Dumpbot as well.

As with Linebot, the threshold values may require some adjustment. Sensor readings can be determined by using the RCX's **View** button. If you are using the NQC program, then set LEFT_THRESHOLD to a value slightly higher than the sensor reading directly over the black line. Set RIGHT_THRESHOLD to a value a little lower than the sensor reading over the blank surface. For example, if the sensor reads 39 over the line and 56 over the blank surface, then try a LEFT_THRESHOLD of 42 and a RIGHT_THRESHOLD of 53. The RCX Code version of the program has three values that may need adjustment. The first value is the highest reading that is considered dark by the sensor watcher (shown as 42 in the illustration). This should be set to the same value as LEFT_THRESHOLD in the NQC program—specifically it should be a little higher than the reading directly over the black line. The second value is the smallest reading that is considered light by the sensor watcher, which should be equal to the previous value plus one (shown as 43 in the illustration). The third value is the high value of the range tested in the **check & choose** block. This should be one less than RIGHT_THRESHOLD—specifically it should be a little lower than the reading over the blank surface. Refer to the previous chapter on Linebot for more information about adjusting the thresholds.

DELIVERING CARGO

Our second program for Dumpbot is a bit more complex. It relies on the improved line-following algorithm, which cannot be implemented in RCX Code; hence only NQC sample code is provided. The goal of the second program is to follow a gray line until it reaches one end, at which point it dumps its cargo. It then turns around and follows the line back to the other end. The ends of the line are indicated by black squares called *stoppers*.

Most of the code for this program has already been written. The linebot3.nqc program contains the follow_line function, and the dumpbot1.nqc program contains the dump function for dropping off cargo. The only new action needed is for Dumpbot to be able to turn around. For this, a turn_around function can be written.

```
void turn_around()
{
    // start turning
    OnRev(LEFT);
    OnFwd(RIGHT);

    // wait until not over line
    until(EYE >= LINE_THRESHOLD);

    // wait until over line again
    until(EYE < LINE_THRESHOLD);
    Off(LEFT+RIGHT);
}
```

In order to turn around, the treads are run in opposite directions. Since Dumpbot is presumably already on the line, it first waits until it can't see a line, then continues turning until it sees the line again. Combining this with the code from linebot3.nqc, dumpbot1.nqc, and a suitable main task, we have the following:

```
// dumpbot2.nqc
//follow a line, dump the cargo, then return

// linebot sensor, motors, and constants
#define EYE      SENSOR_2
#define LEFT    OUT_A
#define RIGHT   OUT_C
#define LINE_THRESHOLD       51
#define STOPPER_THRESHOLD    42
#define TURN_SPEED           3
#define INITIAL_TIME  4
```

```
// dumper motor and constants
#define DUMPER OUT_B
#define DUMP_TIME    70
#define UNDUMP_TIME  100

void dump()
{
    // dump the cargo bin
    OnFwd(DUMPER);
    Wait(DUMP_TIME);
    Rev(DUMPER);
    Wait(UNDUMP_TIME);
    Off(DUMPER);
}

int direction, time, eye, ok_to_stop;

sub follow_line()
{
    // initialize the variables
    direction = 1;
    time = INITIAL_TIME;
    ok_to_stop = 0;

    // start driving
    OnFwd(LEFT+RIGHT);

    while(true)
    {
        // read the sensor value
        eye = EYE;

        if (eye < LINE_THRESHOLD)
        {
            // we are either over the line or a stopper
            if (eye > STOPPER_THRESHOLD)
                ok_to_stop = 1;
            else if (ok_to_stop == 1)
            {
                // found a stopper
                Off(RIGHT+LEFT);
                return;
```

```
                }
            }
        else
        {
            // need to find the line again
            ClearTimer(0);
            if (direction == 1)
            {
                SetPower(RIGHT+LEFT, TURN_SPEED);
                Rev(LEFT);
            }
            else
            {
                SetPower(RIGHT+LEFT, TURN_SPEED);
                Rev(RIGHT);
            }

            while(true)
            {
                // have we found the line?
                if (EYE < LINE_THRESHOLD)
                {
                    time = INITIAL_TIME;
                    break;
                }
                if (Timer(0) > time)
                {
                    // try the other direction
                    direction *= -1;
                    time *= 2;
                    break;
                }
            }

            SetPower(LEFT+RIGHT, OUT_FULL);
            Fwd(RIGHT+LEFT);
        }
    }
}

void turn_around()
```

```
{
    // start turning
    OnRev(LEFT);
    OnFwd(RIGHT);

    // wait until not over line
    until(EYE >=LINE_THRESHOLD);

    // wait until over line again
    until(EYE < LINE_THRESHOLD);
    Off(LEFT+RIGHT);
}

task main()
{
    SetSensor(EYE, SENSOR_LIGHT);

    follow_line();
    dump();
    turn_around();
    follow_line();
}
```

The test surface for this version of Dumpbot should meet the same criteria as for Linebot. Specifically, a gray or green line that is 3/4″ thick on a white surface. Curves should have a 6″ or larger radius, and stoppers should be at least 3/4″ long. Create the path of your choice (I used a large "U" shape) with a stopper at each end. Place Dumpbot near one end with the light sensor close to the line and some suitable cargo in the cargo bin, then run the program. If all goes well, Dumpbot should follow the line to the other end, dump the cargo, turn around, and return to the first end. Like all robots that use the light sensor, some adjustment to the light thresholds (LINE_THRESHOLD and STOPPER_THRESHOLD) may be required. Start with LINE_THRESHOLD set to a value midway between the readings over the blank surface and those directly over the line. Set STOPPER_THRESHOLD to a value midway between the readings for the line and a stopper. Additional hints on how to set and adjust these thresholds were given previously for Linebot.

CONCLUSION

Dumpbot shows how line following can be used as a building block for more sophisticated robots. Even with the simple line-following algorithm, Dumpbot can be extended for more complex behavior. Instead of relying on timing (dumping 5 seconds after the program starts), a bumper could be used to determine the "end of the line." The improved line-following algorithm provides additional flexibility, since Dumpbot then gains the ability to distinguish two different colors on the line.

Dumpbot will be used in chapter 17 as an example of how two robots can communicate with one another.

Chapter 10

Scanbot

Scanbot is a robot that seeks out bright lights. For example, if you stand across the room and point a flashlight at Scanbot, it will head toward you. If you move about the room, it will continue to home in on your location. Scanbot even has a head that turns from side to side as it looks for bright lights, allowing it to proceed on its current course while simultaneously looking for a better alternative.

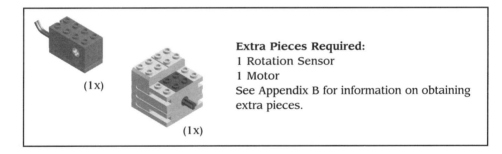

(1x)

Extra Pieces Required:
1 Rotation Sensor
1 Motor
See Appendix B for information on obtaining extra pieces.

(1x)

CONSTRUCTION

Once again, we use the Tankbot chassis (with the original 1:1 gearing). Step 1 shows Tankbot with the addition of a wire on motor port B.

Figure 10–1:
Scanbot Step 1

Scanbot will be using a light sensor as part of a head that can look back and forth searching for light. The head itself is powered by a motor, and the position of the head is determined by a rotation sensor. A frame to support this motor and rotation sensor is built in Steps 2 and 3 using the "wide beam" construction technique. A #6 axle, which will serve as the neck for Scanbot, is also shown. Don't worry, it won't be floating in space for very long.

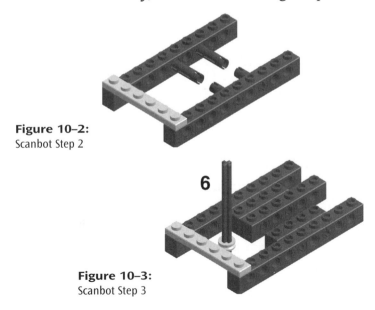

Figure 10–2:
Scanbot Step 2

Figure 10–3:
Scanbot Step 3

The neck is secured in Steps 4 and 5. There should now be two half bushings on the axle, one above the plate and one below. Make sure that these bushings are pushed close enough to the plate to keep the axle from wobbling, but not so close that it cannot spin freely.

Rather than directly gearing the motor to the neck, a belt-and-pulley combination will be used to transfer power. The pulley used here is called a *friction pulley* because of the small rubber ring inside of it. For our purposes we are ignoring its friction capability and using it simply because it is the correct size.

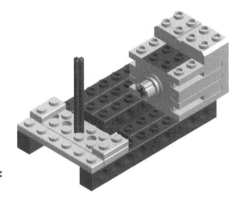

Figure 10–4:
Scanbot Step 4

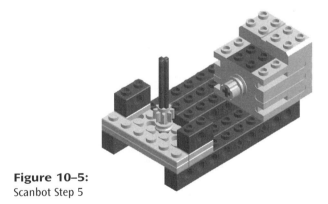

Figure 10–5:
Scanbot Step 5

Steps 6 and 7 show the assembly of the driveshaft using a #8 axle, which passes through the rotation sensor. Two half bushings are used as spacers between the rotation sensor and the worm gear.

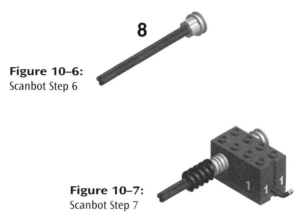

Figure 10–6:
Scanbot Step 6

Figure 10–7:
Scanbot Step 7

In Step 8, the driveshaft is placed on the frame. A white rubber belt should be used to connect the two pulleys (on the motor and on the drive shaft). The worm gear is a little smaller than 2 LEGO units, so it may wiggle back and forth a little. To correct this, slide the worm gear all the way forward against the 1x6 beam, then push the half bushings up tight against the worm gear. The result of this will be a slight space between the bushings and the rotation sensor. As usual, make things tight enough to prevent wiggle, but loose enough so that everything can turn easily. This problem with the worm gear is the reason that two half bushings were used rather than a single regular bushing. Although the total amount of space occupied is the same in both cases, a pair of half bushings has more friction than a regular bushing and will do a better job of holding the worm gear in place.

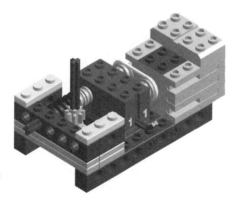

Figure 10–8:
Scanbot Step 8

This drive mechanism accomplishes two very important things: it reduces the rotation speed of the neck (8:1 gear reduction using the worm gear), and it allows slippage if the neck gets stuck. The position of the rotation sensor is critical. The rotation sensor itself has a resolution of 16 ticks per revolution. For Scanbot, we would like more accurate rotation readings, so the sensor is placed before the gear reduction. As a result, the sensor has an effective resolution of 128 ticks per revolution of the neck (16 ticks per revolution times an 8:1 gear reduction). Furthermore, since the intention is for the sensor to give an accurate reading of the neck position, it is important that any slippage occur between the motor and the rotation sensor, rather than between the rotation sensor and the neck (upon which the light sensor is mounted).

The rotation sensor measures relative rotation rather than an absolute rotation. When the program is started, whatever position the sensor is in will be read as the value 0. Rotation in one direction will increment this value, while rotation in the other direction will decrement it. For Scanbot to function, we need to be able to determine if the head is pointing left, right, or straight ahead. This requires knowing an absolute rotation rather than a relative one. This can be accomplished by turning the head to a known position, then resetting the rotation sensor (which sets the value to 0). In Step 9, a small arm is added to the neck, and a $^3/_4$ peg is used as a stopper. We can then turn the neck to a known position simply by running the motor for a few seconds. No matter where the neck originally was, it will eventually bump up against the stopper and remain there. The consequence of this is that we will sometimes be running the motor while the neck is prevented from turning; hence the use of the belt drive to allow slippage.

Another subtle detail in Step 9 is the orientation of the bushing at the top of the neck. The two faces of a bushing are not identical—one is circular, and the other is notched. The notched face fits just perfectly between four "studs" on top of a LEGO piece. We will be taking advantage of this feature later on, so it is important that the notched side of the bushing face upward, as shown below.

A bushing also appears at the end of the drive shaft (the #8 axle with the worm gear on it). This bushing provides a convenient way to manually adjust the neck if necessary. Because the worm gear is a kind of "one-way" mechanism, the neck itself cannot be turned. Instead the worm gear must be turned, which in turn will rotate the neck.

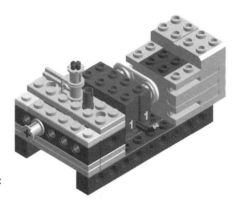

Figure 10–9:
Scanbot Step 9

Figure 10–10:
Scanbot Step 10

Scanbot's head is shown in Step 10. It is built around a light sensor, with a 1x2 beam in front to narrow its field of vision (thus increasing its selectivity).

The head is mounted (upside down) on the neck. It is important to have the correct orientation of the head with respect to the small arm on the neck. When the arm is pointed directly away from the stopper peg, the head should point forward. The sensor wires from the rotation and light sensors should be attached to RCX ports 1 and 3, respectively. The sensor frame then sits on top of the wire connectors from the sensors and motors.

Figure 10–11:
Scanbot Step 11

LOOKING AND STEERING

The basic operation of Scanbot is as follows: drive forward while looking back and forth until a bright light is spotted, then turn a little bit in the direction of the light, and resume looking back and forth.

The rotation sensor can be used to determine the current angle of the head, but since the rotation sensor measures relative rotation, we must first rotate to a known position and reset the sensor. I call this *aligning* the head. A simple way to do this is to turn the head for a fixed amount of time, then stop and reset the sensor. So long as the fixed amount of time is sufficient to move the head from one extreme to another, this will result in alignment of the head. Three seconds should be adequate.

In order to look back and forth, we need to reverse the motor whenever the head has rotated to its rightmost or leftmost rotation. It turns out that the "center" position for the head results in a reading of 53 on the rotation sensor (after the head has been aligned). I chose to have the head sweep an area of 270 degrees (135 degrees both right and left). Since the rotation sensor counts 16 steps per rotation, and the head is geared down by 8:1, a full rotation (360 degrees) would be 128 steps. Thus the leftmost rotation is 48 less than the center position (53 − 48 = 5), and the rightmost rotation is 48 more than the center (53 + 48 = 101). Whenever the rotation sensor reads less than 5, the head has turned too far to the left, and the motor should be set to the forward direction. Whenever the sensor reads greater than 101, the head is too far to the right and the motor should be run backward.

While the head is looking back and forth, the light sensor needs to be monitored. Whenever a bright light is detected, the head is stopped, which results in "freezing" the value of the rotation sensor. This value can then be checked to see if it is to the right or to the left of center. In order to prevent unnecessary zigzagging, a little slop is built into the checks. Any value within 2 of center is considered good enough not to require a turn. The turning itself is quite simple—one tread is reversed for one half a second. A complete implementation of this in NQC is shown below:

```
// scanbot1.nqc
// a simple tracking program for Scanbot

// sensors
#define EYE     SENSOR_3
```

```
#define ANGLE SENSOR_1

// motors
#define LEFT OUT_A
#define HEAD OUT_B
#define RIGHT OUT_C

// other constants
#define SLOP 2

#define CENTER 53
#define SWEEP 48       // 135 degrees

#define ALIGN_TIME    300          // 3 seconds
#define TURN_TIME     50           // 1/2 second

#define THRESHOLD     65

void align()
{
    Rev(HEAD); // begin turning left
    OnFor(HEAD, ALIGN_TIME);
    Off(HEAD);            // stop turning
    Fwd(HEAD);            // prepare for later activation
    ClearSensor(ANGLE); // make this position "0"
}

task main()
{
    // configure sensors
    SetSensor(EYE, SENSOR_LIGHT);
    SetSensor(ANGLE, SENSOR_ROTATION);

    align();

    // look and steer...
    On(LEFT+RIGHT+HEAD);

    // now we can start the steering task
    start steer;

    // keep looking back and forth
    while(true)
    {
```

```
            if (ANGLE < CENTER-SWEEP)
            {
                Fwd(HEAD);
            }
            else if (ANGLE > CENTER+SWEEP)
            {
                Rev(HEAD);
            }
        }
    }

    task steer()
    {
        while(true)
        {
            if (EYE >= THRESHOLD)
            {
                Off(HEAD); // stop looking
                PlaySound(SOUND_CLICK);

                if (ANGLE < CENTER-SLOP)
                {
                    // turn left
                    Rev(LEFT);
                    Wait(TURN_TIME);
                    Fwd(LEFT);
                }
                else if (ANGLE > CENTER+SLOP)
                {
                    // turn right
                    Rev(RIGHT);
                    Wait(TURN_TIME);
                    Fwd(RIGHT);
                }

                On(HEAD);  // resume looking
            }
        }
    }
```

The same behavior can be implemented in RCX Code. By default, the Robotics Invention System software doesn't show the sensor watchers and stack controllers for optional sensors (such as the rotation sensor). To enable these blocks, you must change one of the preferences. From the main menu screen, select "**Getting Started**," then "**Set Up Options**," to get to the preferences screen. At this point the software will try to communicate with the RCX. If a dialog box appears to inform you that the RCX cannot be found, select "**Continue**"—for this particular option the RCX isn't needed. Press the "**Advanced**" button at the bottom of the screen to get to the second page of preferences, then click on the option for "**Unlock rotation sensor watcher**" if it is not already checked. Now you can return to the main menu and build a program that uses the rotation sensor.

We'll start by programming the head to align itself, then rotate back and forth. The RCX Code for this runs under the program stack itself and uses a few *my commands*. The **add to counter** command may seem out of place, but its purpose will be explained later.

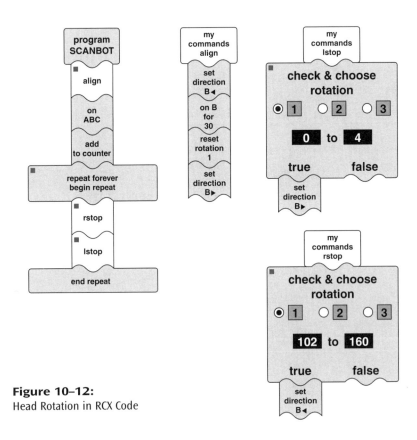

Figure 10–12:
Head Rotation in RCX Code

The code to steer the robot should respond to a light sensor. A first try might look something like this:

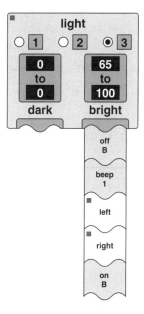

Figure 10–13:
RCX Code Program That Almost Works

The problem is that the sensor watcher will be active even while the head is being aligned. In this case, there's a chance that a bright light will be detected and that the sensor watcher will then stop the head while it attempts to turn. If you get really unlucky, the sensor watcher may even trigger more than once as the head is being aligned. This may prevent the head from rotating all the way to the left, thus preventing true alignment.

What we need is a way of preventing the light sensor from responding until after the head is aligned. This can be done using the counter. When the program starts, the counter is initialized to zero. We can then increment it to 1 after the head has been aligned (thus the **add to counter** command underneath **align**). We then add a **check & choose** for the counter underneath the light sensor watcher, and thus ignore any bright lights during alignment. The revised sensor watcher appears on the next page.

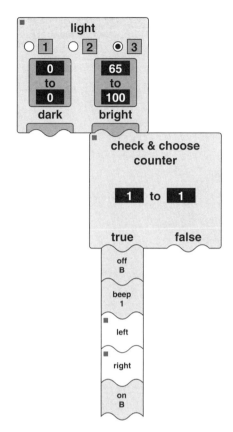

Figure 10–14:
Improved RCX Code for Scanbot

Now you are ready to test Scanbot out. One way is to place a bright light near the floor somewhere in a room, and let Scanbot try to find it. Another option is to aim a flashlight at Scanbot and move around the room (always pointing the flashlight at the robot) while Scanbot tries to catch you.

If the head gets stuck to one side, then it is possible that the wire for motor B is not connected properly. If this happens, then running the motor forward (or backward) will not have the expected effect on the rotation sensor, and Scanbot will become confused. One solution would be to switch the motor directions within the program, but a simpler solution is to ensure that the motor is wired correctly. While the RCX is turned off, manually center the head by turning the bushing at the front end of the driveshaft (the axle that holds the worm gear). Once the head is centered, turn on the RCX and run the Scanbot program. The first thing the program does is attempt to align the head by running the motor backward. This should result in a counterclockwise rotation of the head. If the head turns the other way, then the motor's wiring needs to be reversed. Use a different orientation for the wire as it attaches to either the RCX or the motor itself.

Assuming that the wiring is correct, you are ready to let Scanbot run a little. Scanbot will emit a short beep whenever it has detected a bright light. If this beep is going off too often, then you need to increase the threshold for the light sensor. If Scanbot never detects your light, then the threshold needs to be reduced.

You can also adjust the amount of time that Scanbot turns each time a light is spotted. Smaller values will result in slower, more gradual tracking. Larger values will have a tendency to make Scanbot zigzag as it approaches a light.

IMPROVING THE SCANBOT PROGRAM

Scanbot's program is still rather limited. First of all, the head always takes 3 seconds to align, even if it is already in the left-most rotation. Second, a light threshold must be explicitly set. It would be much nicer if Scanbot followed the *brightest* object, not just any object brighter than a certain threshold. The third problem is that Scanbot turns the same amount whether the light is a little or a lot off center. Programming Scanbot to make slight turns and sharp turns would result in much better tracking. All these issues can be addressed in the NQC version of our program. Unfortunately, RCX Code is too limited to make these adjustments.

Solving the alignment problem is easy. We can use the rotation sensor to determine when the head is no longer moving, and at that point we'll consider the head to be aligned. A revised version of the align function is shown below:

```
void align()
{
    int previous;

    OnRev(HEAD);              // begin turning left
    do
    {
        previous = ANGLE;     // save current position
        Wait(10);
    } while(ANGLE != previous); // if position changed, keep going

    Off(HEAD);
    ClearSensor(ANGLE);       // make this position "0"
}
```

The above function introduces a variable called previous to hold a sensor reading, which is later compared to a more recent reading. This is called *latching* the value. Note that a small delay is necessary between latching the variable and checking it against a second reading. Without the delay, the RCX would take the two sensor readings so close together that the sensor might not change in value even if the head were moving. Now that the head is aligned, we can consider how to use it to scan for a light source.

A simple way would be to start with the head at the leftmost position, then sweep to the right and keep track of the position in which the brightest light was found.

```
int max, level, target;

void scan()
{
    // turn head left
    OnRev(HEAD);
    until(ANGLE <= CENTER-SWEEP);

    // look right
    max = 0;
    Fwd(HEAD);
    while(ANGLE < CENTER+SWEEP)
    {
        level = EYE;
        if (level > max)
        {
            target = ANGLE;
            max = level;
        }
    }
    Off(HEAD);
}
```

There are a couple of subtleties worth pointing out. The reading from the light sensor is latched in the variable level before being compared with, and possibly assigned to, max. The reason for this is that the actual reading of the light sensor may change in the short time between executing the statements that follow. It is important that max get set to the exact same value as was used in the test, otherwise the maximum brightness could be missed.

The second detail is that the rotation sensor is not latched.

Since the head is moving, it sounds like a good idea to latch the angle as well. However, even with latching, the value would still be sampled slightly later than the light sensor, and the extra statement would slow down the loop (increasing the chance of missing a small, bright light). As a compromise, the angle is not latched, but it is recorded before max is updated, minimizing the amount of time between sampling the light sensor and the rotation sensor. Overall, this slight inaccuracy in reading the angle will have little effect.

The code is still not ideal, since each time it is called it may spend up to half its time "rewinding" the head to its leftmost position. What we really want is to be able to call scan when the head is at either the leftmost or rightmost position, and then it will perform its scan either left to right or right to left. Since the code to update the maximum light level (and the angle at which it was found) is the same in both cases, it is put into its own inline function.

```
int target, max;

void check()
{
    int level = EYE;
    if (level > max)
    {
        target = ANGLE;
        max = level;
    }
}

void scan()
{
    max = 0;

    if (ANGLE > CENTER)
    {
        // look left
        OnRev(HEAD);
        while(ANGLE > CENTER-SWEEP)
            check();
    }
    else
    {
        // look right
```

```
            OnFwd(HEAD);
            while(ANGLE < CENTER+SWEEP)
                check();
        }
        Off(HEAD);
    }
```

Now we are ready to address the problem with steering. Instead of a constant amount of turn time, it will be proportional to how far away from center the bright light was.

```
#define SLOP 2
#define TURN 4

void steer()
{
    int time;
    target -= CENTER;

    if (target > SLOP)
    {
        Rev(RIGHT);
        time = TURN * target;
    }
    else if (target < -SLOP)
    {
        Rev(LEFT);
        time = -TURN * target;
    }
    else
        return; // no need to steer, just return now

    // turn for a little while, then go forward again
    Wait(time);
    Fwd(LEFT+RIGHT);
}
```

As in the first Scanbot program, the SLOP value allows Scanbot to continue straight ahead if the target is close to center. The amount of proportional turning can be controlled by the constant TURN. The above code uses a value of 4 for TURN, but you may have to adjust this slightly in your own designs. Smaller values will cause the Scanbot to curve more gently toward the target, whereas larger values can cause Scanbot to overshoot during its turns and oscillate back and forth.

Only a few details, such as initialization and a main loop, remain. The program is shown below:

```
// scanbot2.nqc
// A better Scanbot program

// sensors
#define EYE     SENSOR_3
#define ANGLE   SENSOR_1

// motors
#define LEFT  OUT_A
#define HEAD  OUT_B
#define RIGHT OUT_C

// other constants
#define SLOP 2
#define TURN 4

#define CENTER 53
#define SWEEP 48      // 135 degrees

// variables
int target, max;

void align()
{
    int previous;
    OnRev(HEAD);       // begin turning left

    do
    {
        previous = ANGLE;    // save current position
        Wait(10);
    } while(ANGLE != previous); // if position changed, keep going

    Off(HEAD);
    ClearSensor(ANGLE); // make this position "0"
}

void check()
```

```
    {
        int level = EYE;
        if (level > max)
        {
            target = ANGLE;
            max = level;
        }
    }

    void scan()
    {
        max = 0;

        if (ANGLE > CENTER)
        {
            // look left
            OnRev(HEAD);
            while(ANGLE > CENTER-SWEEP)
                check();
        }
        else
        {
            // look right
            OnFwd(HEAD);
            while(ANGLE < CENTER+SWEEP)
                check();
        }
        Off(HEAD);
    }

    void steer()
    {
        int time;

        target -= CENTER;

        if (target > SLOP)
        {
            Rev(RIGHT);
            time = TURN * target;
        }
        else if (target < -SLOP)
        {
            Rev(LEFT);
```

```
        time = -TURN * target;
    }
    else
        return; // no need to steer, just return now

    // turn for a little while, then go forward again
    Wait(time);
    Fwd(LEFT+RIGHT);
}

task main()
{
    // configure sensors
    SetSensor(EYE, SENSOR_LIGHT);
    SetSensor(ANGLE, SENSOR_ROTATION);

    align();

    // look and steer...
    On(LEFT+RIGHT);
    while(true)
    {
        scan();
        steer();
    }
}
```

USING SCANBOT WHEN SPACE IS LIMITED

Scanbot tends to need a lot of floor space when running the previous program. With some modification we can adapt Scanbot to operate better in crowded environments. There are two basic modifications we need to make. First of all, since Scanbot picks the brightest light source as its target, there is always *some* target available. We can make Scanbot more selective by introducing an *ambient light threshold*. This is just a fancy name for the lowest light sensor reading that Scanbot will consider a valid target. It is very similar to the threshold used in the very first Scanbot program, but rather than programming in a specific value, we'll have Scanbot look around a little bit and pick the value itself. Once this threshold is set, Scanbot will be able to stop if no suitable target is found, rather than simply choosing the next brightest object in the room and homing in on it.

Second, Scanbot is always moving—even when not zeroed in on a light source. We'll change the program so that Scanbot moves forward only when it is locked on target.

The ambient light threshold can be computed by looking from one side to the other and recording the brightest light sensor reading. For a little extra safety, this value should be padded a bit to ensure that Scanbot will pay attention only to lights that are noticeably brighter than the normal background lighting. The following NQC function will compute an acceptable threshold (assuming that the head starts out completely to the left):

```
#define AMBIENT_MARGIN 2

int ambient;

void measure_light()
{
    // determine max ambient light level
    ambient = 0;
    OnFwd(HEAD);
    while(ANGLE < CENTER+SWEEP)
    {
        int level = EYE;
        if (level > ambient) ambient = level;
    }
    Off(HEAD);
    ambient += AMBIENT_MARGIN;
}
```

According to our new strategy, Scanbot should drive forward only when it is locked on target. This requires several modifications to the steer function, as shown below:

```
void steer()
{
    int time;

    if (max <= ambient)
    {
        Off(LEFT + RIGHT);
        return;
    }

    target -= CENTER;

    if (target > SLOP)
```

```
    {
        OnFwd(LEFT);
        OnRev(RIGHT);
        time = TURN * target;
    }
    else if (target < -SLOP)
    {
        OnFwd(RIGHT);
        OnRev(LEFT);
        time = -TURN * target;
    }
    else
    {
        // we're on target...go straight
        OnFwd(LEFT+RIGHT);
        return; // no need to steer, just return now
    }

    Wait(time);
    Off(LEFT+RIGHT);
}
```

Note how the function first checks to see if the maximum light value exceeded the ambient light threshold. If it didn't, then the motors are turned off and the function returns immediately. In this case, Scanbot will sit still and continue scanning until a bright light appears.

In the case of a turn (either right or left), the drive motors are turned on in opposite directions. The function will then wait the appropriate amount of time and stop the motors.

The last case is when Scanbot is heading directly toward the target. In this case the motors are turned on and the function returns immediately. This allows Scanbot to make its next scanning pass while driving toward the target.

All that remains are some minor modifications to the main task so that it calls measure_light at the proper time and doesn't turn on the drive motors initially. The complete program is shown here:

```
// scanbot3.nqc
// Scanbot program with ambient light threshold

// sensors
```

```
#define EYE    SENSOR_3
#define ANGLE SENSOR_1

// motors
#define LEFT  OUT_A
#define HEAD  OUT_B
#define RIGHT OUT_C

// other constants
#define SLOP 2

#define TURN   4
#define CENTER 53
#define SWEEP  48      // 135 degrees

// a little extra range for ambient light #define AMBIENT_MARGIN 2

// variables
int target, max;
int ambient;

void align()
{
    int previous;

    OnRev(HEAD);        // begin turning left

    do
    {
        previous = ANGLE;    // save current position
        Wait(10);
    } while(ANGLE != previous);// if position changed, keep going

    Off(HEAD);
    ClearSensor(ANGLE);       // make this position "0"
}

void measure_light()
{
    // determine max ambient light level
    ambient = 0;
    OnFwd(HEAD);
    while(ANGLE < CENTER+SWEEP)
```

```
        {
            int level = EYE;
            if (level > ambient) ambient = level;
        }
        Off(HEAD);
        ambient += AMBIENT_MARGIN;
    }

void check()
{
    int level = EYE;
    if (level > max)
    {
        target = ANGLE;
        max = level;
    }
}

void scan()
{
    max = 0;

    if (ANGLE > CENTER)
    {
        // look left
        OnRev(HEAD);
        while(ANGLE > CENTER-SWEEP)
                check();
    }
    else
    {
        // look right
        OnFwd(HEAD);
        while(ANGLE < CENTER+SWEEP)
            check();
    }

    Off(HEAD);
}

void steer()
{
    int time;
```

```
        if (max <= ambient)
        {
            Off(LEFT + RIGHT);
            return;
        }

        target -= CENTER;

        if (target > SLOP)
        {
            OnFwd(LEFT);
            OnRev(RIGHT);
            time = TURN * target;
        }
        else if (target < -SLOP)
        {
            OnFwd(RIGHT);
            OnRev(LEFT);
            time = -TURN * target;
        }
        else
        {
            // we're on target...go straight
            OnFwd(LEFT+RIGHT);
            return; // no need to steer, just return now
        }

        Wait(time);
        Off(LEFT+RIGHT);
    }

task main()
{
    // configure sensors
    SetSensor(EYE, SENSOR_LIGHT);
    SetSensor(ANGLE, SENSOR_ROTATION);

    align();
    measure_light();

    // look and steer...
    while(true)
    {
```

```
                scan();
                steer();
        }
    }
```

This program performs rather well in homing in on a flashlight or other bright light source. The only caveat is that the target light needs to be turned off when the program is first started and is determining the ambient light threshold. Once the head makes a complete sweep from left to right, it is then safe to turn on the flashlight (or other light) and let Scanbot home in on it. If you then turn off the light, Scanbot will stop and wait for the light to reappear.

As a side note, the technique of making the program able to set its own light thresholds can be applied to many robots that rely on light sensors for input.

CONCLUSION

Scanbot is a fun robot to play with. Unlike the previous ones, which wandered about randomly, Scanbot seems to have a more directed purpose. The construction of Scanbot required a few new elements: the worm gear and pulley-and-belt drive. It also introduced the rotation sensor, which is perhaps the most versatile MINDSTORMS sensor. With it, a robot can measure speed, distance, position, and other quantities. Because the rotation sensor is not part of the Robotics Invention System, use of it has been minimized in this book. However, once you own a rotation sensor you will certainly find new uses for it in your own robots.

Scanbot is also the last of our robots based on Tankbot. This versatile chassis has served us well for the last several chapters, but it is now time to move on to other vehicle designs such as three-wheeled vehicles and rack and pinion steering. The Tankbot chassis (in the guise of Dumpbot) will make one final appearance in Chapter 17.

Chapter 11
Tribot

So far, all of the robots have been based on the Tankbot chassis, using caterpillar treads for both motion and steering. Tread-based vehicles are compact, able to traverse a variety of terrain, and easy to maneuver. However, treads also have a few disadvantages. First of all, LEGO treads come in only one size, which puts some restrictions on the size and speed of tread-based robots. Second, the treads are a compromise between providing enough traction to propel the vehicle forward and having enough slip to allow turning. Last, the process of the treads moving around the hubs involves a significant amount of friction; hence some of the motors' energy is wasted, thus limiting the speed and power of such robots even more.

This chapter introduces a three-wheeled robot that steers in the same way as the previous tread-based designs, but uses wheels rather than treads. A simple bumper is added to this chassis, and the resulting robot can be thought of as a variation of Bumpbot. Of course, the three-wheeled chassis may also be adapted for other purposes.

THE SWIVEL WHEEL

Tribot is powered by two rear wheels, each driven by its own motor. Steering is accomplished in much the same manner as with the tread-based designs. When both wheels turn at the same speed and in the same direction, the vehicle will be propelled in a straight line forward (or backward). Turning one wheel faster than the other will force Tribot to turn. Very tight turns can be made by running the wheels in opposite directions.

The two wheels alone, however, are not sufficient to balance Tribot—it requires a third support. Attaching this wheel can cause a bit of a problem. When Tribot is moving straight ahead, the wheel should point forward. When turning, this wheel must point in the direction of the turn, otherwise the friction of the skidding wheel would interfere with the turn. However, we don't have to aim the wheel ourselves—all we need to do is allow it to pivot freely, and it will align itself in whichever direction is required.

The key to making this work is to ensure that the pivot axis is offset from the wheel's axle, as shown in Figure 11-1. In such an arrangement, the wheel will always trail slightly behind the pivot axis and tend to line up with the direction of travel. This sort of wheel mechanism can be found on everything from grocery carts to bed frames in the real world.

Figure 11-1:
Swivel Wheel

CONSTRUCTION

Tribot's wheels are attached directly to their respective motors without any intervening gears, so, as with Tankbot, the wheels spin at the rate of the motors themselves. However, since the wheels have a larger diameter and do not suffer from the same internal friction as treads, Tribot moves significantly faster than Tankbot. A pair of wheels are mounted on their hubs and then attached to a pair of motors in Steps 1 and 2. Be sure to use the correct-size wheels—they should be slightly larger than the motor so that they will be able to touch the ground and propel Tribot. The motor assembly can then be set aside for later.

Figure 11-2:
Tribot Step 1

Figure 11-3:
Tribot Step 2

Steps 3 and 4 illustrate the mounting used to hold the swivel wheel in place. The structure is built around a pair of liftarms, a #8 axle, and a #6 axle. The crossblock provides the actual mounting hole for the swivel wheel, and four bushings are used as spacers. Since all of the pieces involved use cross-shaped holes, the entire structure is quite rigid.

Figure 11-4:
Tribot Step 3

Figure 11-5:
Tribot Step 4

The mounting is held between two beams, as shown in Step 5, and the structure continues to be built up in Steps 6 and 7. The oval-shaped pieces added in Step 6 are another kind of liftarm: a 1×3 liftarm. They provide a very compact way to add vertical bracing to a structure. In Step 7 a pair of #3 axles are inserted through beams into the tops of the 1×3 liftarms, thus completing the vertical bracing.

Figure 11-6:
Tribot Step 5

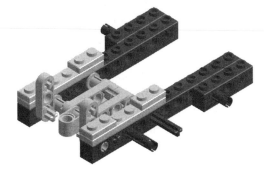

Figure 11-7:
Tribot Step 6

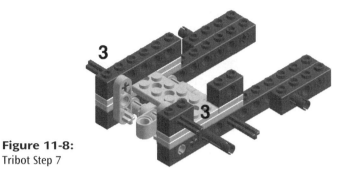

Figure 11-8:
Tribot Step 7

The swivel wheel is started in Step 8. The new piece here is called a #1 *angle brick*. When we combine it with the #5 angle brick in Step 9, we get a very compact swivel wheel. The #6 axle extending upward serves as the swivel's axis, and it is inserted as shown in Step 10. The swivel-wheel portion of our robot is now complete, but we still need to build a bumper on the front end.

Figure 11-9:
Tribot Step 8

Figure 11-10:
Tribot Step 9

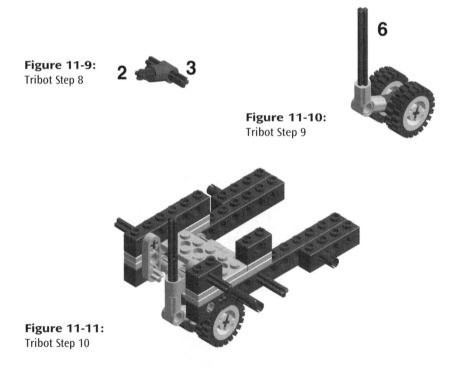

Figure 11-11:
Tribot Step 10

The bumper consists of a pair of angle beams and a pair of #8 axles, joined in the center by an *axle extender*. The axle extender is often used to construct a long axle out of two shorter ones. In this case, we effectively get a #16 axle by combining two #8 axles. The bushings, as usual, are used as spacers. In order to operate smoothly, the bumper needs to be "square" (the axles must be perpendicular to the angle beam). Without the spacers, manual adjustment would be required to keep things square. A small black rubber band is used to give the bumper the proper tension. One minor detail is the addition of a friction peg on the bumper's angle beam (just above where the rubber band hooks on). When the bumper hits an obstacle, it pivots such that one end of the angle beam comes into contact with the touch sensor. The friction peg helps keep the touch sensor pressed while the angle beam continues to pivot past the sensor. Overall, this bumper design extends a little farther forward than previous designs, which is a good thing since Tribot moves faster than our other robots and needs more time to react.

The swivel wheel's axle is secured in place by a series of bushings and the 2×6 technic plate. This plate may seem like overkill, but its function is actually quite important. Part of Tribot's weight must be supported by the swivel wheel. In the orientation shown here, the axle is in front of the actual wheels. This means that the downward force of Tribot's weight will make the axle tilt; the top will be pushed forward while the bottom is pushed backward. This tilting creates additional friction between the axle and the crossblock it is mounted through, thus hampering its swivel ability. Adding the top plate solves this problem quite nicely.

Figure 11-12:
Tribot Step 11

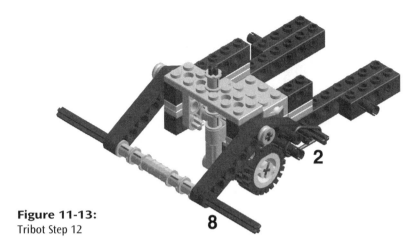

Figure 11-13:
Tribot Step 12

The bumper, motor assembly, and RCX are put together in Step 13. The RCX is secured to the bumper assembly with a pair of 1×6 beams, and the motors are mounted at the back of the RCX. Finally, the wiring for the motors and touch sensor may be completed.

Figure 11-14:
Tribot Step 13

PROGRAMMING

To test out Tribot, we will write a simple program to drive forward until it hits something, then back up a little, turn a little, and resume. A sample program in NQC appears below. By now, its structure should be familiar. Sensors, motors, and constants are defined at the top, and the main task has an infinite loop that waits for a condition to be true (the bumper being hit), then reacts accordingly (back up for 0.50 second, spin around for 0.20 second).

```
// tribot.nqc
// three wheeled robot

// motors and sensors
#define LEFT OUT_A
#define RIGHT OUT_C
#define BUMP SENSOR_1

// constants
#define REV_TIME      50
#define TURN_TIME     20

task main()
{
    SetSensor(BUMP, SENSOR_TOUCH);

    On(LEFT+RIGHT);

    while(true)
    {
        until(BUMP==1);

        // back up a little
        Rev(LEFT+RIGHT);
        Wait(REV_TIME);

        // spin a little
        Off(LEFT);
        Wait(TURN_TIME);

        // resume going forward
        OnFwd(LEFT+RIGHT);
    }
}
```

A similar program written using RCX Code is shown below.

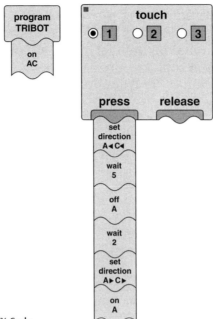

Figure 11-15:
Tribot Program in RCX Code

VARIATIONS

The Tribot chassis can be adapted for any of the previous Tankbot-based robots. The bumper itself can be easily modified to provide separate left and right bumpers, each with its own touch sensors. The resulting vehicle could then be programmed much the same way as Bugbot.

For some tasks, such as line following, a slower robot is required. The three-wheeled design can still be used, but some gear reduction between the motors and the wheels is required. A 3:1 gear reduction will result in a very powerful robot that still moves at a reasonable speed.

If a faster robot is needed, the wheels can be replaced by larger ones, such as the large motorcycle-style wheels included in the Robotics Invention System set. The motors will have to work rather hard in such a design, and they may even stall on rough surfaces such as carpet. Top speed on a smooth floor, however, is quite impressive.

A variation of Tribot that uses only a single motor for driving and turning is presented in the next chapter.

Both the three-wheeled and tread designs rely on evenly

matched motors for straight travel. If the motors (or the wheels or treads) vary slightly between the left and the right sides, there will be a tendency for the vehicle to curve slightly instead of traveling in a straight line. An alternative drive mechanism that avoids this problem is presented in chapter 14.

Chapter 12

Onebot

The RCX only has three outputs, and this can become a serious limitation when designing a robot. In our examples so far, two motors have been used just for movement, leaving only a single output free for other functions. What if those other functions need two independent motors? Onebot is a variation of Tribot that uses a single motor to move around. Although Onebot itself is only a simple bumper car, its design can serve as the basis for more complicated robots.

DRIVING WITH ONE MOTOR

The previous robots were very flexible in their driving and steering. They could go forward or backward, turn left or right (in forward and reverse), or even spin in place. Although convenient, not all of these operations are needed for most tasks. For example, the combination of driving straight ahead and turning in reverse is sufficient for many tasks.

Forward motion in Tribot was accomplished by turning both motors on in the forward direction. Turning was accomplished by stopping the left motor and running the right motor in the reverse direction. (Tribot also traveled straight backward, but we are ignoring that mode for Onebot.) As you can see, the right motor must be able to turn in both directions, but the left motor only needs to turn forward. The ratchet mechanism discussed in chapter 4 is perfect for allowing motion in one direction while preventing motion in the other direction.

The most obvious way of using a single motor to power both wheels would be to put both wheels on a single axle, then use the motor to turn that axle (via gears or pulleys). Unfortunately, if a ratchet is then used to prevent the left wheel from reversing, it will also prevent the right wheel from reversing. The wheels must therefore both have their own axle, but the problem remains of how to transfer power from the motor to both axles. If gears are used, the ratchet will still prevent the right wheel from turning, since it will "lock up" the entire gearing mechanism (including the motor itself). Pulleys and belts can be used instead to allow one side of the drive mechanism to slip while the other side still turns. Although functional, the use of belts limits the amount of power that can be delivered to the wheels, as well as the overall efficiency of the drive mechanism.

A more elegant solution is to use the differential. As discussed in chapter 4, the differential consists of a shell and two independent axles. One axle may turn faster or slower than the other (they may even turn in opposite directions), but their average will always equal the speed of the shell itself. If we use the motor to turn the shell, then the default case is for both axles to turn at the same speed. Reversing the motor will result in trying to reverse both axles. Stopping one of the axles with a ratchet will result in the other wheel's turning twice as fast (so that the average still equals the speed of the shell).

This is the mechanism that is used to power Onebot. It is like the ratchet splitter presented in Chapter 4, with one important difference. The ratchet splitter uses ratchets on both axles, while Onebot only has a single ratchet. This is because the ratchet splitter was designed to turn one axle or the other, but never both at the same time. In contrast, Onebot always needs to turn one axle; it is only the second one that is sometimes powered and sometimes stopped.

CONSTRUCTION

Onebot uses the same swivel-wheel and bumper assembly as Tribot. If you still have Tribot around, remove the RCX and rear motors. Otherwise, follow the instructions in Chapter 11, Steps 3 through 12. The resulting assembly is shown on the top of the next page.

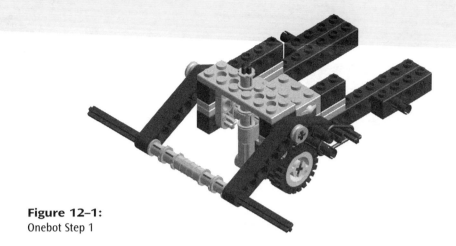

Figure 12–1:
Onebot Step 1

The previous section outlined the principles behind Onebot's drive mechanism, but there are also a few implementation details of its construction that are worth pointing out. The differential and related gears are built in Steps 2 through 5. Note that each side has its own #6 axle, which is attached to a bevel gear within the differential shell. A third bevel gear inside the differential transfers power between these two axles.

A stack of 1x1 plates is placed behind the crown gear (a 1x1 brick could have been used instead, but the Robotics Invention System set does not include any). This construction provides firm support behind the crown gear. Without it, the crown gear would have a tendency to flex slightly and not mesh properly when under strain. The result would be an annoying clicking sound and a robot that didn't move very well.

Figure 12–2:
Onebot Step 2

Figure 12–3:
Onebot Step 3

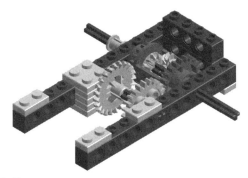

Figure 12–4:
Onebot Step 4

Figure 12–5:
Onebot Step 5

The motor and ratchet are added in Steps 6 through 8. An 8-tooth gear is placed on the motor, which then meshes with the crown gear (24 teeth), resulting in a 3:1 gear reduction. Tribot was able to propel itself effectively without the use of gear reduction, but it was also using two motors. In Onebot, the power from a single motor must be divided between both wheels. A gear reduction is thus used to increase the power to the wheels (at the expense of speed).

The motor itself is mounted on special mounting brackets. The brackets are attached above and below the 1x12 beams. The motor is then slid into place and secured by a 1x4 plate. This results in a compact yet robust mounting for the motor. There are quite a few situations where these brackets can be used to position motors in places where traditional building techniques would fail.

The ratchet design is fairly simple: gravity pulls the crossblock down such that it engages with the teeth of the 16-tooth gear. Try turning the axle clockwise; the gear's teeth catch the crossblock at an angle and get locked in place. Now try turning the axle slowly in the counterclockwise direction; the teeth are now able to push the crossblock up and out of the way. If you give the axle a quick turn counterclockwise you'll notice that the crossblock gets "kicked up" a bit. If you turn it quickly enough, the crossblock will flip all the way around to the other side. This potential problem will be solved later on by placing the RCX so that it prevents the crossblock from getting kicked up too high.

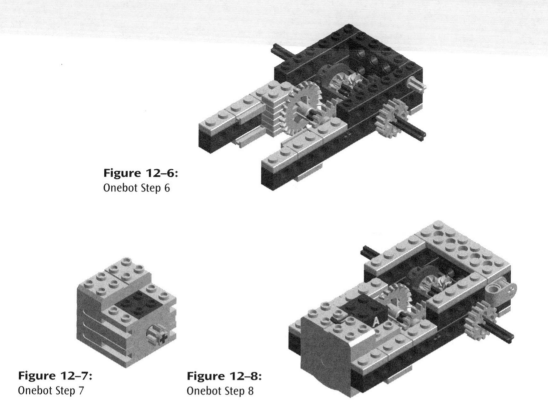

Figure 12–6:
Onebot Step 6

Figure 12–7:
Onebot Step 7

Figure 12–8:
Onebot Step 8

The final steps for Onebot involve attaching the front bumper assembly to the drive mechanism, then adding the RCX and wires.

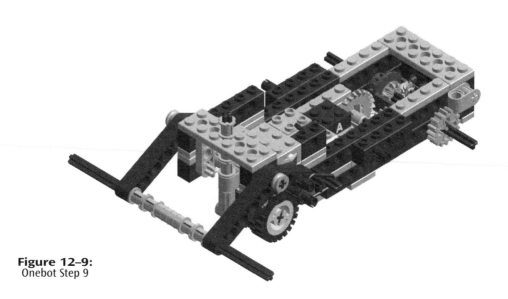

Figure 12–9:
Onebot Step 9

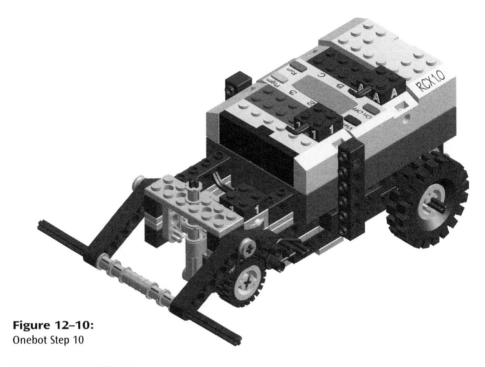

Figure 12–10:
Onebot Step 10

PROGRAMMING

The program for Onebot is similar to Tribot's program. The differences are that only a single motor is controlled and that there is no reverse mode—just forward and reverse/turn. A sample NQC program follows:

```
// onebot.nqc
// robot driven by a single motor

// motor and sensor
#define DRIVE OUT_A
#define BUMP SENSOR_1

// constant
#define REV_TIME     40

task main
{
    SetSensor(BUMP, SENSOR_TOUCH);

    On(DRIVE);
```

```
while(true)
{
    until(BUMP==1);

    // back up / turn a little
    Rev(DRIVE);
    Wait(REV_TIME);

    // resume going forwards
    Fwd(DRIVE);
}
}
```

A similar program for RCX Code is shown below:

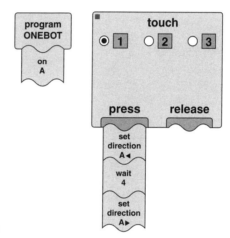

Figure 12–11:
Onebot Program in RCX Code

CONCLUSION

Although somewhat limited in overall movement, the single-motor design behind Onebot can be useful when both other motors are required for auxiliary functions. The drive mechanism can also be adapted to use different wheels or even treads.

Chapter 13

Steerbot

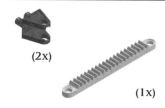

	Extra Pieces Required:
(2x)	2 Steering gear bearings
	1 20-tooth rack (1x10)
(1x)	See Appendix B for information on obtaining extra pieces.

RACK AND PINION STEERING

Many of the previous robots have been based on the tread-driven chassis introduced with Tankbot. Although such designs are easy to build and program, they are limited by the size and traction of the treads themselves. This puts a practical limitation on the size and speed of robots that can be built with the standard LEGO treads.

In the real world a different steering mechanism, known as *rack and pinion steering*, is much more common. In fact, it is the same mechanism used in virtually all automobiles. Rack and pinion steering provides complete flexibility on the shape and size of tires, thus it is often a good choice for vehicles that do not perform adequately with tracks. In the case of Steerbot, the steering will be done by the front wheels, and the drive train will power the rear wheels. Other combinations are also possible—forklifts often steer the rear wheels, sport utility vehicles power all 4 wheels, and some cars even provide 4-wheel steering—however, the basic principles are the same.

When driving straight, a vehicle's wheels are parallel to one another. In order to turn, the front wheels are angled slightly. In a sense, this angle aims the front of the vehicle to the right or left. As long as the wheels remain angled, the vehicle will continue to turn.

In order for the wheels not to slide, the outside wheels must turn slightly faster than the inside wheels. This is the same effect that we exploited in the track-based designs, but this time cause and effect are reversed. Whereas before we changed the speed of one side in order to cause a turn, now the turn (which is caused by the angle of the wheels) will cause a difference in speed to be required. Even if we were willing to use 2 motors (1 for the right-rear wheel, and 1 for the left-rear wheel), the required difference in speed depends on the sharpness of the turn and is rather complicated.

The solution involves using a *differential*, which was described in chapter 4. We can use a motor to turn the shell of the differential. The differential functions by distributing power to both wheels in such a way as to minimize the overall friction. When driving straight, this means that both wheels will turn at the same speed. When turning, however, the inside wheel will turn slightly slower than the differential shell, and the outside wheel will turn slightly faster. This variation in speed will happen without any direct intervention—the only thing we need to do is spin the differential itself.

CHASSIS CONSTRUCTION

Steerbot is mechanically the most complicated robot so far, but if you look carefully you will see the same familiar techniques employed on earlier robots. The only truly unique portion of Steerbot is the rack and pinion steering. Building a rack and pinion steering mechanism from scratch can be complicated (and bulky). Fortunately, there are several specialized LEGO pieces which make this task much simpler. Figure 13-1 shows these special pieces and how they combine to form a rack and pinion steering mechanism.

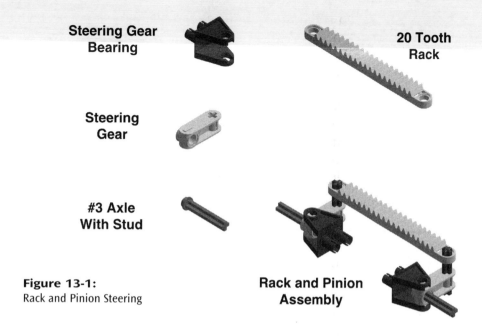

Steering Gear Bearing

20 Tooth Rack

Steering Gear

#3 Axle With Stud

Rack and Pinion Assembly

Figure 13-1:
Rack and Pinion Steering

To assemble the mechanism, insert a *#3 axle with stud* through a *steering gear* (also known as a 1x3 crossblock). The steering gear then snaps into a *steering gear bearing*, which allows the steering gear to pivot back and forth. A standard axle, such as a #2 or #3 axle, should then be inserted in the back of the steering gear. The 20-tooth rack can then be used to join a pair of these steering assemblies. As the rack moves back and forth it will cause the steering gears to pivot appropriately.

Steps 1, 2, and 3 begin building up Steerbot's chassis, including a differential and the special brackets used to mount a motor. The new pieces near the front of the frame (left side of the illustration) are something of a "hack." This is the area where the rack will need to slide back and forth; thus a smooth surface is required. Ideally, a couple of *tiles* (plates without studs on top) would be used. Since the Robotics Invention System doesn't include any regular tiles, these odd pieces are used instead.

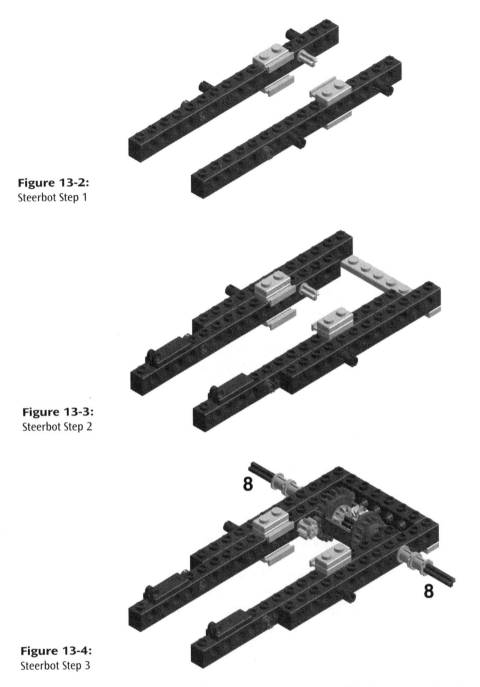

Figure 13-2:
Steerbot Step 1

Figure 13-3:
Steerbot Step 2

Figure 13-4:
Steerbot Step 3

The motor (with a crown gear attached) is slid into position in Step 4, then locked in place in Step 5. The crown gear is a 24-tooth gear with curved teeth that allows it to mesh at a right angle with another gear (see chapter 4 for a more detailed explanation of the crown gear). The gearing between the motor and differential may appear a bit odd. The motor turns a crown gear (24 teeth), which meshes with an 8-tooth gear for a 3:1 reduction. This 8-tooth gear then meshes with the 24-tooth side of the differential for a 1:3 gear ratio. The resulting ratio is simply 1:1. So why not skip the 8-tooth gear and mesh the crown gear directly with the differential? In order to do this, the crown gear would need to be moved closer to the differential. Unfortunately, at such close proximity, the center of the crown gear would collide with the shell of the differential. By introducing the 8-tooth gear, space is added between the crown gear and the differential without altering the overall speed and power (aside from a minor amount of friction).

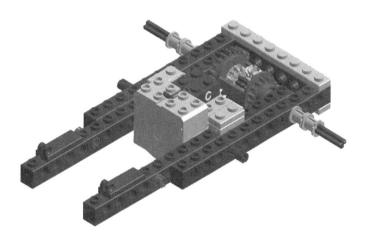

Figure 13-5:
Steerbot Step 4

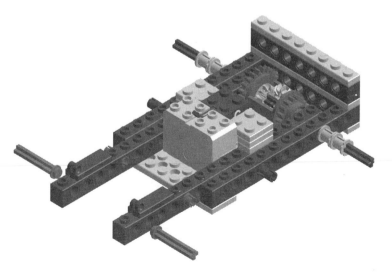

Figure 13-6:
Steerbot Step 5

A touch sensor is used to determine when the steering is centered. Steps 6 and 7 build this sensor as well as completing the front axle assemblies for the steering mechanism. A pair of #3 axles are inserted through the back end of the steering gears (a.k.a. 1x3 crossblocks), and half bushings are placed as spacers.

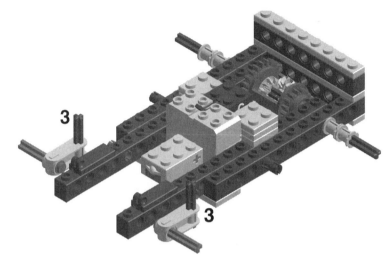

Figure 13-7:
Steerbot Step 6

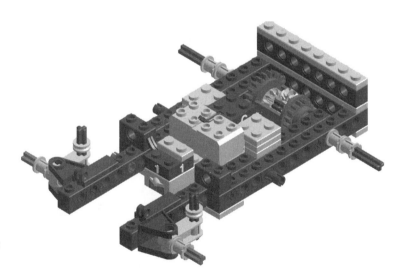

Figure 13-8:
Steerbot Step 7

Normally, the rack and pinion steering would be completed simply by adding the rack. However, in our case we need to add a few pieces below the rack so that the touch sensor gets pressed when the rack is centered. This construction is shown in Steps 8 and 9, after which it is added to the chassis in Step 10. A 1x2 cross beam (the green one) with a #2 axle is used to activate the touch sensor. The bottom of the 20-tooth rack is hollow, thus a small piece (such as a 1x2 plate) cannot be attached directly to it. Instead, a 1x8 plate is used below the rack to hold the 1x2 plate (and cross beam) in place. Observe how the wheels rotate right and left as the rack is moved back and forth and how the touch sensor gets pressed only when the wheels are centered (or very close to center).

Figure 13-9:
Steerbot Step 8

Figure 13-10:
Steerbot Step 9

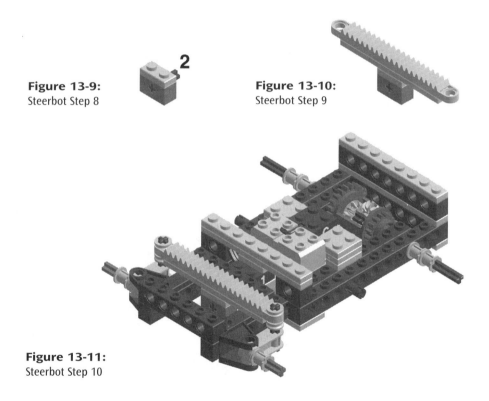

Figure 13-11:
Steerbot Step 10

The steering is powered by a motor as shown in Steps 11, 12, and 13. A pair of pulleys and a blue rubber belt are used to transfer power from the motor. The difference in size between the pulleys provides a reduction of approximately 7:2, which provides a good compromise between power and speed. A belt was intentionally used (as opposed to gears) in order to allow slippage when the steering has been turned completely to the left or right. As usual, support for the motor must take into account the rather unusual shape of its bottom.

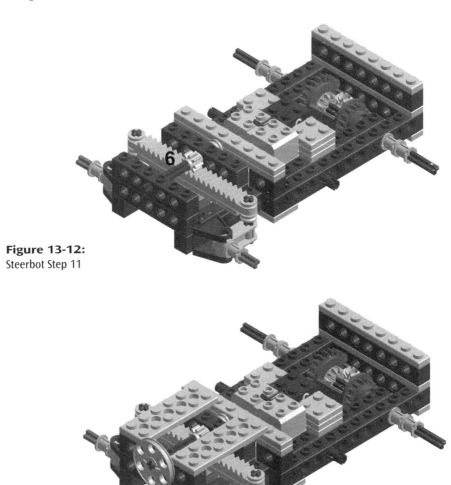

Figure 13-12:
Steerbot Step 11

Figure 13-13:
Steerbot Step 12

Figure 13-14:
Steerbot Step 13

The bumper uses a #10 axle, a #8 axle, and a pair of angle beams for its basic structure as shown in Steps 14, 15, and 16. The small black rubber band gives this bumper a decidedly "springy" feel, but otherwise it is very similar to the previous bumper designs. In most cases, the tension on the rubber band will keep it in place. However, if you find that it falls off, simply add a 2x2 plate above the rubber band to hold it in place.

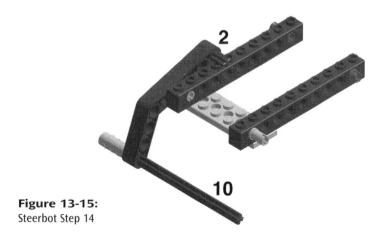

Figure 13-15:
Steerbot Step 14

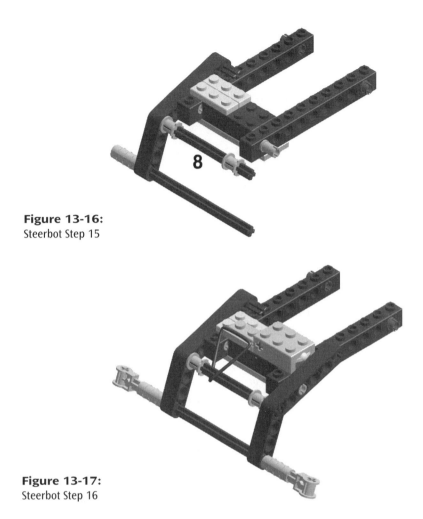

Figure 13-16:
Steerbot Step 15

Figure 13-17:
Steerbot Step 16

Steerbot is completed by adding the bumper, RCX, and wheels to the chassis. Unlike all of the previous robots, the RCX faces backward. This is because the motor would tend to block the IR port if the RCX were facing forward.

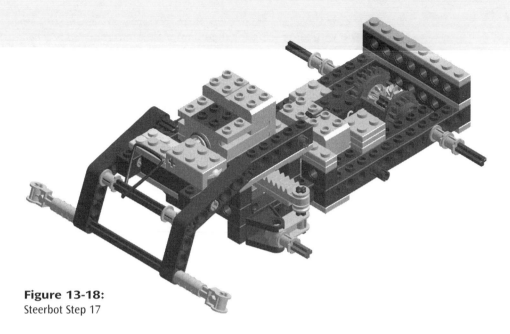

Figure 13-18:
Steerbot Step 17

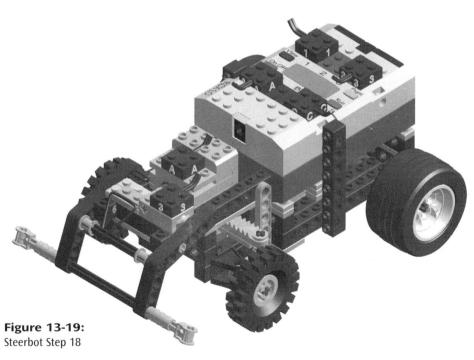

Figure 13-19:
Steerbot Step 18

> The vertical bracing for the bumper uses *1x5 rounded half beams* rather than standard 1x6 beams, and the reason for this is somewhat obscure. Personally, I don't like the "chunky" tires used in front. I much prefer the wide tires used in the rear, so I originally built Steerbot using four wide tires. These wide tires would rub against a 1x6 if it was used for bracing, therefore the 1x5 rounded half-beams were used instead. Of course the Robotics Invention System only has two of the wide tires, so most people will be using the chunky tires in front. However, if you happen to have four of the wide tires in hand, use them—Steerbot will look much nicer.

PROGRAMMING

We will first introduce three operations to manage the steering. These operations, named left, right, and center, turn the wheels to the left, right, or center positions. Sample code for both RCX Code and NQC is shown below:

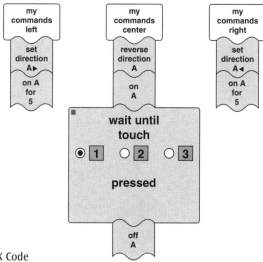

Figure 13-20:
Steering Commands in RCX Code

```
#define CENTER  SENSOR_1
#define STEER   OUT_A

#define FULL_TURN_TIME        50

void left()
{
    Fwd(STEER);
    OnFor(STEER, FULL_TURN_TIME);
}
```

```
void right()
{
    Rev(STEER);
    OnFor(STEER, FULL_TURN_TIME);
}

void center()
{
    Toggle(STEER);
    On(STEER);
    until(CENTER==1);
    Off(STEER);
}
```

As usual, the NQC code begins with defining names for the sensors, motors, and constants to be used. In all other aspects, the RCX Code and NQC versions are quite similar. To turn the wheels left, we run the steering motor forward for a little while—1/2 second to be exact. This number was chosen because it provides ample time for the steering to be moved from the far right extreme to the far left extreme. Turning the wheels right is the same, except the motor is run in the reverse direction.

Centering the steering is a little different. The center operation assumes that the steering had previously been moved with the right or left operation. As a result, switching the motor's direction will get the steering moving back toward center. The motor remains on until the centering touch sensor is triggered. If you find that the center operation is turning either too little or too much, then it may be adjusted by either adding a delay after the sensor is triggered (turning too little), or reversing the motor's direction for a bit (turning too much). Examples of this in NQC are shown below:

```
// version of center () that turns a little extra
void center()
{
    Toggle(STEER);
    On(STEER);
    until(CENTER==1);
    Wait(10);
    Off(STEER);
}
```

```
// version of center() that turns a little less
void center()
{
    Toggle(STEER);
    On(STEER);
    until(CENTER==1);
    Toggle(STEER);
    Wait(10);
    Off(STEER);
}
```

From these steering primitives, we can build the RCX Code or NQC programs shown below:

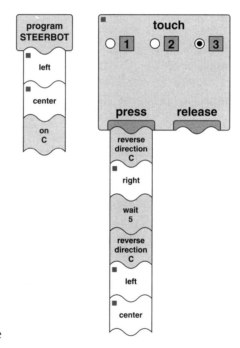

Figure 13-21:
Steering Program in RCX Code

```
// steerbot1.nqc
// a simple bumper car with rack & pinion steering

#define CENTER        SENSOR_1
#define BUMP          SENSOR_3

#define STEER   OUT_A
```

```
#define DRIVE   OUT_C

#define FULL_TURN_TIME 50
#define STRAIGHT_TIME   50

void left()
{
    Fwd(STEER);
    OnFor(STEER, FULL_TURN_TIME);
}

void right()
{
    Rev(STEER);
    OnFor(STEER, FULL_TURN_TIME);
}

void center()
{
    Toggle(STEER);
    On(STEER);
    until(CENTER==1);
    Off(STEER);
}

task main()
{
    SetSensor(CENTER, SENSOR_TOUCH);
    SetSensor(BUMP, SENSOR_TOUCH);

    // align wheels
    left();
    center();

    // drive forward
    On(DRIVE);

    while(true)
    {
        // wait for bump
        until(BUMP==1);

        // back up while turning
        Rev(DRIVE);
```

```
        right();
        Wait(STRAIGHT_TIME);

        // forward while turning other direction
        Fwd(DRIVE);
        left();

        // center steer
        center();
    }
}
```

The NQC program starts by configuring the sensors, but otherwise the programs are similar. The wheels are initially centered by first turning completely to the left, and then calling the center operation, at which point Steerbot begins driving forward. When it hits an obstacle, it will back up while turning to the right, resume moving forward while turning a bit to the left, then continuing straight ahead.

One common problem is that the #2 axle underneath the rack isn't sticking out far enough to activate the touch sensor. If this is case, then the program will get stuck during the center operation and Steerbot won't go anywhere. To correct this, adjust the #2 axle such that it makes good contact with the touch sensor.

Improper alignment of the bumper is another potential problem. If the angle beams used in the bumper are not lined up straight, then they may not activate the touch sensor and Steerbot will never detect an obstacle; it will bash into a wall and continue trying to go forward. This can be corrected by adjusting the front ends of the angle beams (where they attach to the #10 axle) so that the angle beams remain nice and straight.

If you find that the bumper falls off after repeated collisions, then consider adding additional bracing to hold it in place.

ASYNCHRONOUS STEERING

In the previous example, when the program wanted to turn right it called the right operation, which didn't return until the wheels had turned completely to the right. This sort of behavior, where an operation does not return until some external operation is complete, is known as *synchronous* behavior. Sometimes it is desirable to be able to start an operation, then continue doing other processing while the operation completes. This execution model is called *asynchronous* behavior.

Consider the process of cooking a "bacon and eggs" breakfast. One way to do this would be first to cook the bacon, and only when it was finished move on to preparing the eggs. In this case the bacon is being cooked synchronously. Another option would be to start cooking the bacon, then begin preparing the eggs while the bacon was still cooking. In this case the bacon is being cooked asynchronously. Preparing the eggs while still keeping an eye on the bacon is a bit more difficult than simply cooking one then the other, but overall it takes less time. This tradeoff often applies to software as well. Synchronous behavior is often the simplest to implement, but in some cases asynchronous behavior is worth the extra effort.

Since turning the wheels completely to the right or left can take up to a $1/2$ second, steering control is a good candidate for asynchronous operation.

> The following programming technique cannot be implemented within RCX Code; only NQC sample code will be provided.

The key to asynchronous behavior (at least in the RCX) is to use multiple tasks. We will create a separate task whose sole function is to turn the wheels to a desired position. Of course, other tasks (known as *clients*) will need some way of interacting with the steering task in order to specify which position the steering mechanism should be set to. It is also desirable for these clients to be able to determine when the steering task has completed its work.

This is starting to sound pretty complicated, and one of the techniques programmers use to make complicated systems easier to understand is called *abstraction*. The key to abstraction is to take some feature of the system and separate "how to use it" from "how it works." The "how to use it" portion is called the *interface* to the feature, and "how it works" is the *implementation*.

We make such abstractions in our everyday lives as well. We constantly use devices—televisions, microwave ovens, automobiles—without completely understanding how they actually work. In many cases the internal operation of these devices has changed considerably over the years, yet the way they are used remains relatively unchanged. Of course, new features are added over time, but typically in a way that is compatible with the old features. Abstracting and designing good programming interfaces, which can grow to accommodate new features while still remaining compatible with existing clients, is one of the most challenging tasks a programmer must face.

Returning to asynchronous steering, the way a client interacts with the steering task is the *interface* to the steering feature. As mentioned above, the interface must allow a client to specify which position to set the steering to and then determine when the operation has been completed. In NQC we build this interface using several constants and a pair of variables as shown below:

```
// interface for steering task
#define STEER_LEFT      -1
#define STEER_CENTER    0
#define STEER_RIGHT     1

int steer_goal;              // set this to steer
int steer_current = -999;    // check this
```

A client may indicate its desired steering position by setting the value of steer_goal to one of the three position constants (such as STEER_LEFT). The task may check to see what the current steering position is by reading the value of steer_current. Initially, steer_current is set to an invalid number (-999) to indicate that the steering task has not been initialized. Once the steering task completes its initialization, this variable will contain a valid steering position.

Note that a client does not need to know anything about how the steering task works (its implementation). It is sufficient to understand only the interface to the task.

The implementation of our steering task is a bit more complicated than its interface. A complete listing is shown here:

```
// implementation of steering task
#define CENTER          SENSOR_1
#define STEER           OUT_A
#define FULL_TURN_TIME      50

task steer_task()
{
    SetSensor(CENTER,  SENSOR_TOUCH);

    // align the wheels
```

```
    OnFwd(STEER);
    Wait(FULL_TURN_TIME);
    Toggle(STEER);
    until(CENTER==1);
    Off(STEER);

    // set to our initial position
    steer_goal = STEER_CENTER;
    steer_current = STEER_CENTER;

    while(true)
    {
        int goal;

        goal = steer_goal;
        if (goal != steer_current)
        {
            if (goal == STEER_LEFT)
            {
                // turn left
                Fwd(STEER);
                OnFor(STEER, FULL_TURN_TIME);
            }
            else if (goal == STEER_RIGHT)
            {
                // turn right
                Rev(STEER);
                OnFor(STEER, FULL_TURN_TIME);
            }
            else
            {
                Toggle(STEER);
                On(STEER);
                until(CENTER==1);
                Off(STEER);
            }

            steer_current = goal;
        }
    }
}
```

As usual, the code begins with some defines for the motor, sensor, and constants. The task itself starts by configuring the sensor, then centering the wheels using code that is almost identical to that from our earlier synchronous example. After the wheels are centered, the two interface variables, steer_goal and steer_current, are set to indicate that the wheels are centered.

The task then enters an infinite loop which checks to see if the steer_goal differs from steer_current. The actual value of steer_goal is copied into the local variable goal, prior to this test. This is because the steering task must make a number of decisions, and a change of goal in the midst of these decisions could lead to unpredictable behavior. Since a client may change steer_goal at any time, the steering task must read it once, then make all decisions from that single reading. Each time through the loop, this copy is refreshed with the latest value from steer_goal.

When steer_goal does not match steer_current, the task then takes the appropriate action (turning the wheels to the left, right, or center), then updates steer_current with the new location. Although the steering happens asynchronously with respect to the clients, each steering operation must run to completion before another one can be acted upon. For example, if steer_goal is set to STEER_RIGHT, the steering task will begin to move the wheels to the right. This may not complete for another 1/2 second. Meanwhile a client may set steer_goal to STEER_LEFT. The steering task will continue moving the wheels all the way to the right before checking steer_goal again and determining that the wheels must be moved back to the left. In short, once a steering operation has started, it cannot be canceled.

Before the steering task may be used, the client program must start it. In addition, it is often desirable for the client to wait until the steering task has finished its initialization before moving on to other things. These 2 steps can be done with the following code:

```
start steer_task;
until(steer_current == 0);
```

Using the asynchronous steering task, we can write a new version of our bumper car program. In this version we will add a new behavior: the robot will beep as it backs up. The following code snippet will spend $1\frac{1}{2}$ seconds beeping every $\frac{1}{2}$ second.

```
repeat (3)
{
    PlayTone(880, 30);
    Wait(50);
}
```

If we used this code with the previous synchronous steering functions, then the beeps wouldn't start until the steering had been completed. Now that we have asynchronous steering, the beeps can start immediately. The entire program follows.

```
// steerbot2.nqc
// multi-tasked steering


//  interface for steering task
#define STEER_LEFT      -1
#define STEER_CENTER     0
#define STEER_RIGHT      1

int steer_goal;         // set this to steer
int steer_current = -999;           // check this

// implementation of steering task
#define CENTER          SENSOR_1
#define STEER           OUT_A
#define FULL_TURN_TIME      50

task steer_task()
{
    SetSensor(CENTER, SENSOR_TOUCH);

    // align
    OnFwd(STEER);
    Wait(FULL_TURN_TIME);
    Toggle(STEER);
    until(CENTER==1);
    Off(STEER);

    steer_goal = 0;
    steer_current = 0;
```

```
        while(true)
        {
            int goal;

            goal = steer_goal;
            if (goal ! = steer_current)
            {
                if (goal == STEER_LEFT)
                {
                    // turn left
                    Fwd(STEER);
                    OnFor(STEER, FULL_TURN_TIME);
                }
                else if (goal == STEER_RIGHT)
                {
                    // turn right
                    Rev(STEER);
                    OnFor(STEER, FULL_TURN_TIME);
                }
                else
                {
                    Toggle(STEER);
                    On(STEER);
                    until(CENTER==1);
                    Off(STEER);
                }

                steer_current = goal
            }
        }
}

// main task...

#define BUMP          SENSOR_3
#define DRIVE         OUT_C
#define STRAIGHT_TIME 50
```

```
task main()
{
    SetSensor(BUMP, SENSOR_TOUCH);

    start steer_task;
    until(steer_current == 0);

    On(DRIVE);

    while(true)
    {
        until(BUMP==1);
        Rev(DRIVE);
        steer_goal = STEER_RIGHT;
        repeat (3)
        {
            PlayTone(880, 30);
            Wait(50);
        }

        steer_goal = STEER_CENTER;
        Fwd(DRIVE);
    }
}
```

If synchronous behavior is desired while using the steering task, then the client needs to wait for steer_current to equal the desired setting before moving on. For example, in the above program, the operation of centering the wheels before driving forward could be made synchronous by replacing:

```
steer_goal = STEER_CENTER;
Fwd(DRIVE);
```

with:

```
steer_goal = STEER_CENTER;
until(steer_current == STEER_CENTER);
Fwd(DRIVE);
```

VARIATIONS

Rack and pinion steering is not limited to bumper cars. In general, it can be used with the sensor mechanisms from the previous robots to make faster versions of Bugbot, Scanbot, and others.

The touch sensor isn't a very reliable way to determine when the wheels are centered, and as a result Steerbot tends to drift a little to the left or right. A more precise steering mechanism can be constructed by using a rotation sensor. For improved accuracy, there should be a gear reduction between the rotation sensor and the 8-tooth gear that turns the rack. That way the rotation sensor will be able to measure smaller amounts of movement of the rack. A belt drive, if used, should be between the motor and the rotation sensor, and not between the rotation sensor and the rack. If a rotation sensor is used, then the value of the sensor itself can replace the `steer_current` variable in the asynchronous steering task.

Chapter 14

Diffbot

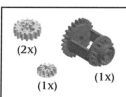

Many of the robots so far have been powered by 2 independent motors, each driving one side of the robot. Whether treads (Tankbot) or wheels (Tribot) are used, the principle remains the same: drive both sides at the same speed to go straight, and drive one side faster than the other to turn. Since driving straight requires both motors to be turning at the exact same speed, slight variations in the motors can cause the robot to drift quite a bit to the left or right.

Diffbot is still driven by treads, but avoids the drifting problem through the use of 2 differentials. Instead of operating autonomously, Diffbot will be controlled directly by a remote control that is built from the RCX and sensors. Both the drive mechanism and the remote control may be incorporated into other robot designs.

(2x)

(1x)   **(1x)**

Extra Pieces Required:
2 16 tooth gears
1 differential
1 bevel gear (only required when using set #9747)
See Appendix B for information on obtaining pieces.

DUAL-DIFFERENTIAL DRIVE

Diffbot uses a special drive mechanism constructed from 2 differentials. The easiest way to understand how it works is to start building it as shown in Steps 1, 2, and 3.

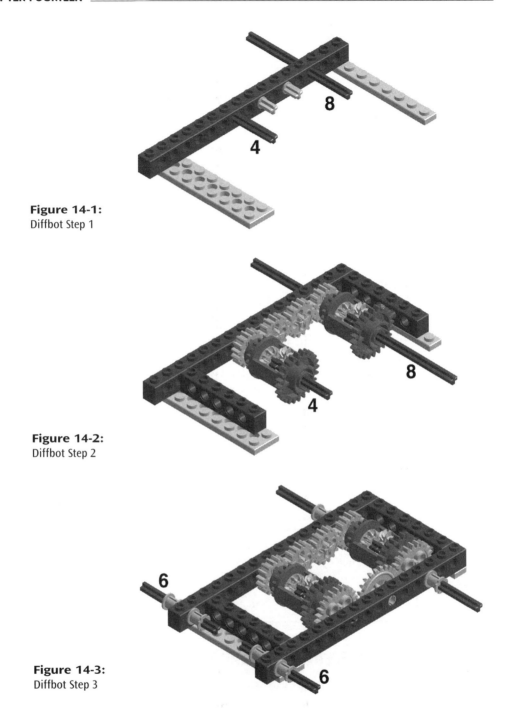

Figure 14-1:
Diffbot Step 1

Figure 14-2:
Diffbot Step 2

Figure 14-3:
Diffbot Step 3

The rear differential, hereafter referred to as the *primary* differential, drives axles that will eventually power the treads. This is the traditional application of a differential: power is applied to the shell, and the axles will then turn so that their average speed is that of the shell itself. This gives us an easy way to power both axles from a single motor, thus avoiding the drift associated with slight variations in drive systems with 2 motors. The secondary differential (roughly in the middle of the chassis) is the key to turning. The far axles of both differentials are joined by a series of 16-tooth gears. Since there is an even number of gears (an odd number of meshes) on that side, the axle for the secondary differential will always turn in the opposite direction of the axle turning the far tread. The near axles of the differentials are also linked together, but this time using three gears. Since this is an odd number of gears, both axles will turn in the same direction. Ideally, all three of these gears would be ordinary 24-tooth gears; however, the Robotics Invention System only includes 4 such gears and we would need a total of 5 for Diffbot. As a workaround, a crown gear (which also has 24 teeth) is used as the middle gear. If you have an extra 24-tooth gear (perhaps from service pack #5229), it may be used instead.

Consider the case of both treads turning at the same speed. Both axles of the primary differential will turn at the same speed as the shell itself. The axles of the secondary differential will also be spinning. The near axle will turn in the same direction as the primary differential, while the far axle will turn in the opposite direction. Since these axles are turning at the same speed but in opposite directions, the shell of the second differential will remain motionless. In effect, the shell of the second differential turns at a speed that is *one half of the difference* between the speeds of the right and left treads. The rotation of the differentials can be summarized as follows:

If the primary differential is stopped, then the treads will turn

$$primary \quad = \frac{left + right}{2}$$

$$secondary = \frac{left - right}{2}$$

at the speed of the secondary differential, but in opposite directions. This results in the vehicle turning in place.

If both differentials are turned at the same speed in the same direction, then the right tread will be stopped while the left one turns at twice the speed of the differentials. The reverse case happens when the differentials are turned at the same speed, but in opposite directions. This can be used for very sharp turns, but since turning in place is already possible, perhaps we can arrange for more gentle turning.

If the second differential is turned at a fraction of the speed of the primary, then both treads will still turn in the same direction, but one will be turning slightly faster than the other. The greater the ratio between the two differentials, the more gradual the turn will be. For Diffbot, a 9:1 gear reduction will be used. Steps 5 through 8 show the gearing used to turn the secondary differential.

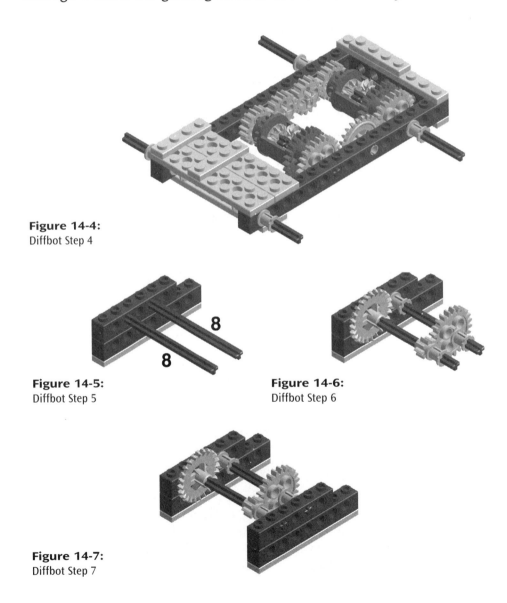

Figure 14-4:
Diffbot Step 4

Figure 14-5:
Diffbot Step 5

Figure 14-6:
Diffbot Step 6

Figure 14-7:
Diffbot Step 7

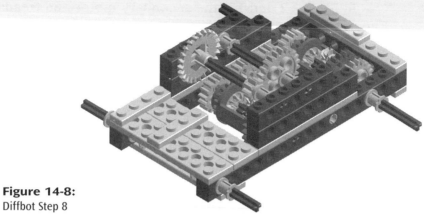

Figure 14-8:
Diffbot Step 8

In Steps 9, 10, and 11, the primary motor is suspended above the differential, using motor brackets without which such a placement would be impossible.

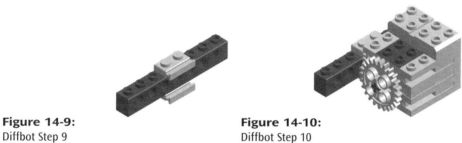

Figure 14-9:
Diffbot Step 9

Figure 14-10:
Diffbot Step 10

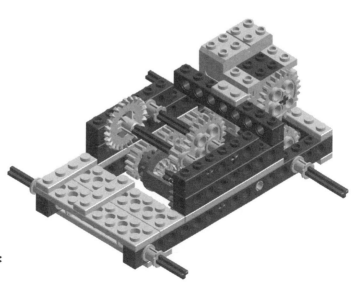

Figure 14-11:
Diffbot Step 11

In Steps 12 and 13, the secondary motor and treads are added. The secondary drive uses a crown gear, which is well supported from behind in order to prevent slipping. A long wire should be connected to each motor as shown. In the next section, these wires will be connected to the remote control for Diffbot.

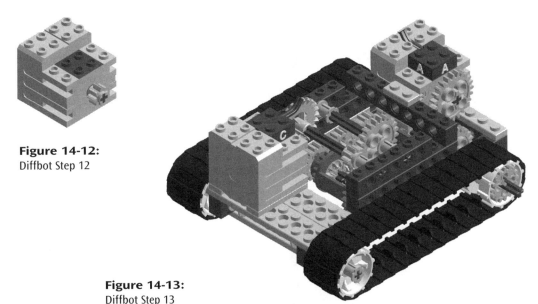

Figure 14-12:
Diffbot Step 12

Figure 14-13:
Diffbot Step 13

CONSTRUCTING THE REMOTE

The remote must be able to control two motors. Furthermore, it is desirable for each motor to be independently run forward, reverse, or turned off. Using two touch sensors to control a single motor is simple. The first touch sensor, while pressed, can be used to run the motor in the forward direction. The second touch sensor can be used to run it in the reverse direction.

It is a little more difficult to use a light sensor to control a motor. Since the motor may be in one of three states (forward, off, reverse), we need three discernable colors that can be placed in front of the light sensor: yellow, green, and black work well. A simple control stick is constructed which moves one of the three colors in front of the light sensor. A rubber band is used to keep the control stick centered when not in use.

Since the controls are built directly above the RCX input connectors, the first step is to attach the sensor wires to the RCX itself as shown in Step 1. Use short wires for ports 1 and 2, and the light sensor for port 3.

Figure 14-14:
Remote Step 1

Figure 14-15:
Remote Step 2

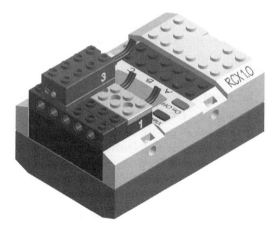

Figure 14-16:
Remote Step 3

The two touch sensors are added and connected to the RCX as shown in Step 4. A double peg is also inserted into the 1x6 beam. This double peg will be used later to anchor the rubber band that centers the control stick.

Figure 14-17:
Remote Step 4

As mentioned before, the control stick uses different colors to trigger a response from the light sensor. In Step 5, a yellow brick, green cross beam, and black brick are used as the foundation for the control stick. A black rubber band should also be placed around the #2 axle protruding from the cross beam.

Figure 14-18:
Remote Step 5

Figure 14-19:
Remote Step 6

The completed control stick is attached to the rest of the control in Step 7. The rubber band is used to provide tension so that the control stick centers itself when not in use. For the right amount of tension the band should be wrapped a second time around the double peg. The wires connected to the RCX's outputs are the other ends of the wires used in Diffbot. The wire from the primary motor (the one with the 24-tooth gear) should be connected to output A, and the secondary motor should be connected to output C.

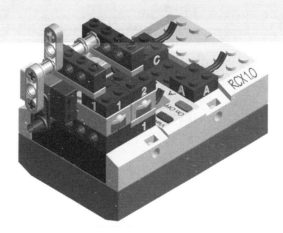

Figure 14-20:
Remote Step 7

PROGRAMMING THE REMOTE

The remote has two touch sensors and a light sensor, each of which requires its own sensor watcher in RCX Code or task in NQC. For the touch sensors, operation is simple: turn on the primary motor (either forward or backward) when the button is pressed, and turn it off when released. RCX Code program stacks for this appear below:

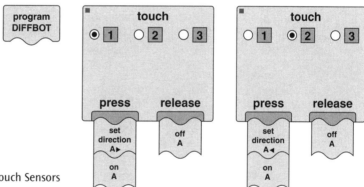

Figure 14-21:
Motor Control with Touch Sensors
in RCX Code

The light sensor must control both forward and reverse direction of the secondary motor. In RCX Code, a light sensor watcher can define both "light" and "dark" ranges. No action is taken for values between the light and dark ranges. In my case, the sensor registered a value of 44 for black, 50 for green, and 59 for yellow. A suitable cutoff for black is thus 47 (halfway between 44 and 50), and a good cutoff for yellow is 55. Your sensor readings may vary, so be sure to check the values (using the **View** button) and set the cutoffs appropriately. When the black brick is in front of the light sensor, it will register a dark value. The program will then start the motor in the forward direction, wait for a lighter sensor value, then stop the motor. As a result, the motor will run only as long as the black brick is in front of the light sensor. The yellow brick will trigger the other stack, which runs the motor in the reverse direction until darker values are read.

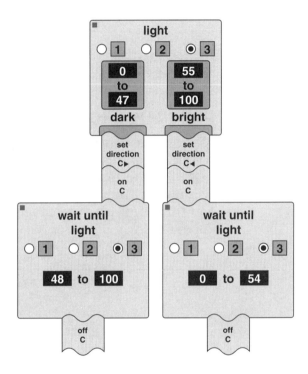

Figure 14-22:
Motor Control with a Light Sensor in RCX Code

A similar program can be written in NQC. Unlike previous programs, which did most of their work within the main task, this one has three auxiliary tasks (fwd_task, rev_task, and steer_task) which do all of the work. As a result, the main task only has to configure the sensors and start the tasks. The resulting code is very similar in structure to the RCX Code program presented above.

```
// diffbot.nqc
// remote control for Diffbot

// motors and sensors
#define DRIVE        OUT_A
```

```
#define TURN         OUT_C
#define FWD_BUTTON   SENSOR_1
#define REV_BUTTON   SENSOR_2
#define STEER        SENSOR_3

// thresholds for steering control
#define LEFT    47
#define RIGHT   55

task main()
{
    // configure sensors
    SetSensor(FWD_BUTTON, SENSOR_TOUCH);
    SetSensor(REV_BUTTON, SENSOR_TOUCH);
    SetSensor(STEER, SENSOR_LIGHT);

    // start all tasks
    start steer_task;
    start fwd_task;
    start rev_task;

    // this task is done now...
}

task fwd_task()
{
    while(true)
    {
        // drive forward while button is pressed
        until(FWD_BUTTON == 1);
        OnFwd(DRIVE);
        until(FWD_BUTTON == 0);
        Off(DRIVE);
    }
}

task rev_task()
{
    while(true)
    {
        // drive backwards while button is pressed
        until(REV_BUTTON == 1);
        OnRev(DRIVE);
        until(REV_BUTTON == 0);
```

```
        Off(DRIVE);
    }
}

task steer_task()
{
    while(true)
    {
        // should we turn left?
        if (STEER <= LEFT)
        {
            OnFwd(TURN);
            until(STEER > LEFT);
            Off(TURN);
        }

        // should we turn right?
        if (STEER >= RIGHT)
        {
            OnRev(TURN);
            until(STEER < RIGHT);
            Off(TURN);
        }
    }
}
```

Once the remote control is programmed, you're ready to take Diffbot for a spin. Run the program, then start playing with the controls. The touch sensors can be used to drive forward or backward, and the control stick can be used to turn in place. Experiment with turning in place (control stick only) and gradual turns (touch sensor + control stick).

VARIATIONS

The dual-differential drive mechanism used in Diffbot is a good substitute for the Tankbot chassis used in many of the previous robots. Although it is a little larger and more complex to build, the elimination of drift and addition of gradual turning make this drive mechanism a good overall choice as long as space isn't at a premium.

The remote control itself is also a useful construction. The programming can be adapted for other drive mechanisms, and provides a convenient way to control two motors during construction and experimentation with new robots.

Chapter 15
Bricksorter

Brick Sorter is a robot that sorts LEGO bricks by color. It was designed to accommodate the green and yellow 1x2 bricks included in the original Robotics Invention System (set #9719). If you are using version 1.5 of the set (#9747), then green 1x2 cross beams may be substituted for the green 1x2 bricks. The design could easily be adapted for other brick sizes or colors.

Brick Sorter consists of a chute with a light sensor at the bottom. The bricks to be sorted are lined up within the chute. The light sensor determines the color of a brick, and then an ejector arm pushes the brick to either the left or the right. The next brick slides into place and the process is repeated. A touch sensor is used to determine the position of the ejector arm and thus control when it should change direction or stop.

CONSTRUCTION

In order for the bricks to slide down the chute a smooth surface is required. In many LEGO designs, this is accomplished by using tiles (plates with a smooth top). The Robotics Invention System does not include any tiles, so we need a different approach. Turning a pair of beams upside down solves this problem nicely. Friction pegs and double-pegs attach these inverted beams to the rest of the chute, as shown in Steps 1 through 4.

Figure 15-1:
Brick Sorter Step 1

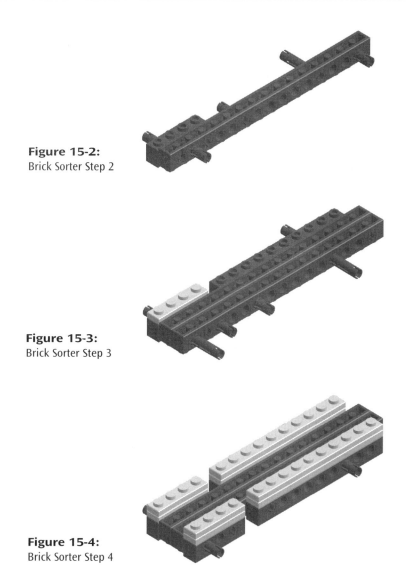

Figure 15-2:
Brick Sorter Step 2

Figure 15-3:
Brick Sorter Step 3

Figure 15-4:
Brick Sorter Step 4

The chute is also designed to minimize the amount of stray light that can enter it and hit the light sensor. This is important because the light sensor not only determines what color brick is in front of it, but also must decide when the chute is empty. This means that the sensor must be able to distinguish between a green brick and no bricks, which in turn means that the chute should be very dark when empty. Step 5 shows how the upper portion of the chute is enclosed in solid bricks (rather than beams).

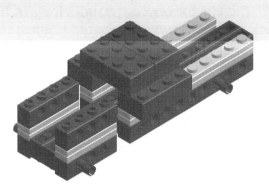

Figure 15-5:
Brick Sorter Step 5

The light sensor needs to be mounted at the end of the chute. It would be easiest to turn the light sensor upside down and attach it directly to the inverted beams. Unfortunately, this would place the sensing portion of the sensor too high to detect the bricks in the chute. Instead, plates are attached to the sensor, as shown in Step 6. Then, the plates are attached to the sides of the chute (Step 7), thus suspending the light sensor from above. The ejector arm needs to slide back and forth so that, just as in Steerbot, a pair of odd hinge pieces are used as substitutes for actual tiles.

Figure 15-6:
Brick Sorter Step 6

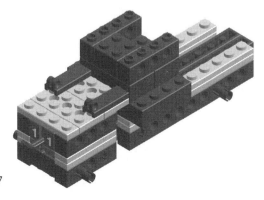

Figure 15-7:
Brick Sorter Step 7

The ejector arm is 16 units long. A beam would be a little too thick, so plates are used instead. Since there are no 1x16 plates, 1x6 and 1x10 plates are combined, as shown in Steps 8 and 9. Note how the seam in the second layer of plates is offset from the seam in the first layer. This is a common construction technique in both the real world and the LEGO world, and it results in stronger structures because there is never a single seam through more than one layer. The arm is completed in Step 10. The three 10-tooth racks combine to make a single long rack.

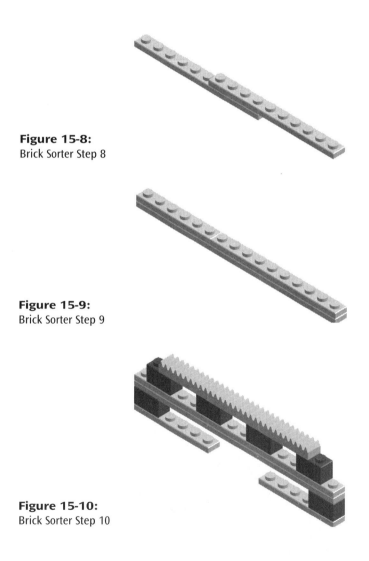

Figure 15-8:
Brick Sorter Step 8

Figure 15-9:
Brick Sorter Step 9

Figure 15-10:
Brick Sorter Step 10

The ejector arm is added to the chute, along with a motor and a touch sensor, in Step 11. Note how the touch sensor detects the spaces between the 1x2 bricks on the ejector arm. This allows the brick sorter to determine the position of the arm.

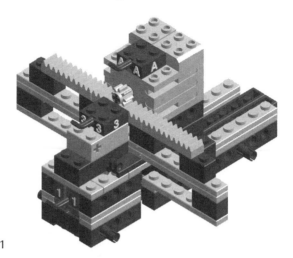

Figure 15-11:
Brick Sorter Step 11

The base for the Brick Sorter is constructed in Steps 12 through 14. Its "T" shape is the key to its stability. Equally important is the fact that there is a layer of plates below the beams making up the base. Later on, vertical beams will be added as legs for the chute. When beams are attached to one another this way, the end of the vertical beam extends slightly below the bottom of the horizontal beam. Without the bottom layer of plates, the brick sorter would rest on the ends of the legs rather than on the base itself. Note that you will need to use yellow 2x4 plates when building the base, since there are not enough of the gray plates.

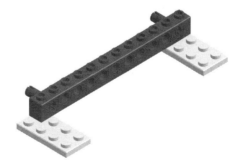

Figure 15-12:
Brick Sorter Step 12

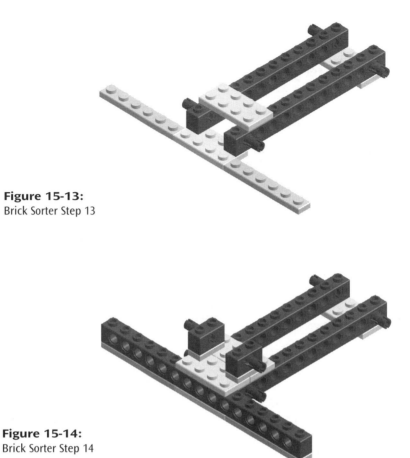

Figure 15-13:
Brick Sorter Step 13

Figure 15-14:
Brick Sorter Step 14

The chute is mounted on a support frame that holds it at just the right angle. If the angle is not steep enough, the bricks will not slide down reliably. If it is too steep, the bricks may fall over as they slide, and the sorter will become jammed. A pair of 1x8 beams are used in front, and 1x12 beams are used in the rear. The resulting angle isn't a natural right triangle (such as 3-4-5), so the vertical "legs" are not quite parallel, and only the front pair of legs uses vertical bracing to assure that they are perpendicular to the floor.

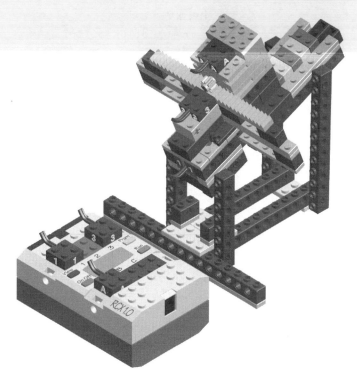

Figure 15-15:
Brick Sorter Step 15

PROGRAMMING

The theory behind Brick Sorter's operation is use of the light sensor to determine if there are bricks in the chute and, if so, the color of the next brick. Depending on the color of the brick (yellow or green), the motor is set to the forward or backward direction. The motor is then run until the touch sensor detects the gap at either end of the arm. At that point, the arm will have already pushed the brick out to one side or the other. The motor's direction is then reversed. Once the center gap is detected, the motor is turned off. After a slight delay (to allow the next brick to fall into place), the process is repeated. An NQC program for running the sorter is shown on the following page.

```
// sorter.nqc
// program for Brick Sorter

#define EYE          SENSOR_1
#define NOTCH        SENSOR_3

#define ARM    OUT_A

// you may need to adjust these
#define NO_BRICKS     46
#define BRIGHT_BRICK  51
#define BRICK_DELAY   10

task main()
{
    // configure sensors
    SetSensor(EYE, SENSOR_LIGHT);
    SetSensor(NOTCH, SENSOR_TOUCH);

    // run until no more bricks
    while(EYE >= NO_BRICKS)
    {
        // check color of brick
        if (EYE > BRIGHT_BRICK)
            Fwd(ARM);
        else
            Rev(ARM);

        // wait until arm has moved to end
        On(ARM);
        until(NOTCH == 1);
        until(NOTCH == 0);

        // reverse direction arm and wait until centered
        Toggle(ARM);
        until(NOTCH==1);
        until(NOTCH==0);
```

```
// stop arm
Off(ARM);

// wait a little while for next brick to slide down
Wait(BRICK_DELAY);
    }
}
```

As you can see, the bulk of the program is contained within a loop that uses the light sensor to determine if more bricks remain to be sorted. A pair of until statements are used to detect the gaps. The first until waits until the touch sensor is pressed, which means it is no longer in the current gap. The second until waits until the sensor is released, which means the next gap has been reached. Note how the Toggle command is used to reverse the direction of the motor regardless of its current direction.

There are two thresholds that need to be determined in the program: BRIGHT_BRICK and NO_BRICKS. To determine the BRIGHT_BRICK threshold, view the light sensor's values when a yellow brick and when a green brick are in front of it. The BRIGHT_BRICK threshold should be about halfway between these two values. The NO_BRICKS threshold should be halfway between the value for a green brick and the value when the chute is empty. If the empty chute's value is too bright, try turning the sorter so that no outside light shines directly into the chute.

The sorter program can also be written in RCX Code. The program's operation requires the nesting of several stack controllers, so once again we create *my commands* for each one of the stack controllers that needs to be nested, then call these commands from within a loop in the main program stack. The threshold values discussed above become the cutoffs in the **repeat while** and **check & choose** stack controllers.

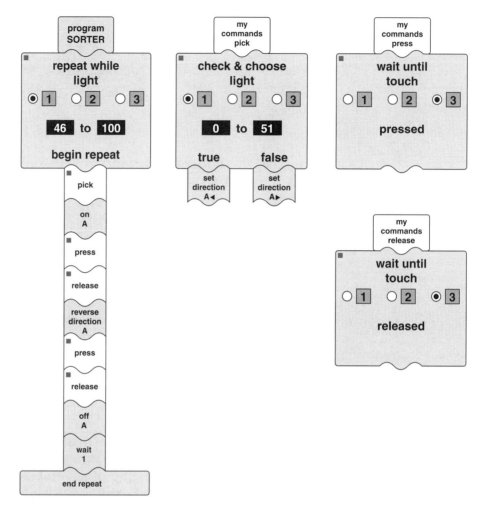

Figure 15-16:
Brick Sorting in RCX Code

SORTING

The sorter is now ready to be used. Carefully slide a few green and yellow 1x2 bricks or cross beams into the chute (if a brick falls on its side, the sorter will become jammed). Make sure the ejector arm is centered, then run the program. If the bricks are not lined up straight or the ejector arm isn't properly centered, Brick Sorter will tend to self-destruct as it attempts to sort the bricks. Special care should be taken to ensure that the ejector arm itself is solidly assembled and that its ends can slide freely in and out of the chute.

If the thresholds are set properly, the robot will stop after a few moments with a pile of green bricks on one side, and a pile of yellow bricks on the other. If it cannot tell green from yellow, then the BRIGHT_BRICK threshold (or the cutoff in the RCX Code **check & choose** block) needs to be adjusted. If the sorter fails to stop when the chute is empty, then you need either to position it so that no external light is shining into the chute, or perhaps to adjust the NO_BRICKS threshold (or the cutoff in the RCX Code **repeat while** block).

CONCLUSION

Many of the previous robots were general-purpose vehicles that could be used as building blocks for a variety of other robots, just the way Linebot was used as a starting point for Dumpbot. By contrast, Brick Sorter is a special purpose robot suitable for little besides sorting some 1x2 bricks. Adapting it for bricks of other sizes or colors would be simple enough, but otherwise its use is fairly limited. The construction and programming techniques, however, are very reusable. For example, using inverted beams to provide a smooth surface is utilized in the next chapter.

Chapter 16
Vending Machine

Vending Machine dispenses small candies, such as M&Ms®. It is composed of two mechanisms: a card reader and a candy dispenser. Special cards are inserted into the card reader, and the appropriate number of candies is then pushed out of the dispenser.

CONSTRUCTION

The cards for Vending Machine are based on a 2x10 plate, with 1x1 plates representing the value of the card. The front right corner of the card must always have a 1x1 plate. Up to 3 more 1x1 plates may be placed along the right edge subject to the following conditions: no two 1x1s may be placed adjacent to one another, and the back 2 rows must be left blank. Two cards (with values 4 and 2) are shown below. The front of the cards is toward the top left corner of the illustration.

Figure 16-1:
Vending Machine Cards

The candy dispenser is similar to the chute constructed for Brick Sorter. Once again inverted beams are used to provide a smooth surface, and double pegs are used to join these beams to the rest of the assembly. Steps 1 through 5 show the construction of the bottom portion of the dispenser. The candy will slide down the dispenser until it hits the 1x2 beam at the end. The 1x10 plates serve as guardrails to keep the candy from accidentally rolling off the side until the dispenser pushes them out.

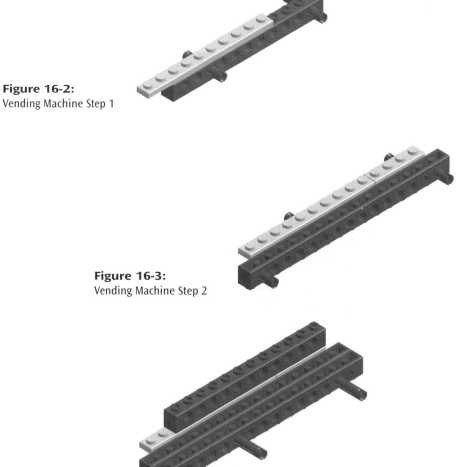

Figure 16-2:
Vending Machine Step 1

Figure 16-3:
Vending Machine Step 2

Figure 16-4:
Vending Machine Step 3

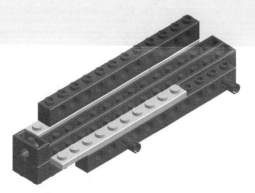

Figure 16-5:
Vending Machine Step 4

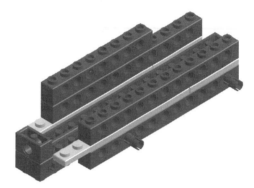

Figure 16-6:
Vending Machine Step 5

The dispenser must be able to push the candy out one at a time. For this, a small arm is turned by a motor—each revolution of the arm will push a single candy out of the dispenser. A touch sensor is used to determine the position of the arm, thus allowing Vending Machine to count how many candies have been dispensed. The beams added in Step 6 provide support for the motor and touch sensor. The arm is added to a #12 axle in Step 7. Note how a crossblock has been placed at the other end of the axle. When the arm is pointed straight down, the crossblock will point up and brush against the touch sensor.

Figure 16-7:
Vending Machine Step 6

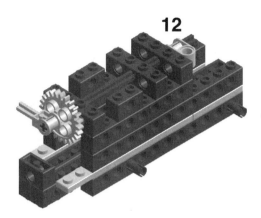

Figure 16-8:
Vending Machine Step 7

The dispenser is completed by adding the motor and touch sensor in Step 8. The arm is turned using a 3:1 gear reduction. Without a gear reduction, the arm would turn much faster: shooting the candies out rather than pushing them. It would also be more difficult for the RCX to respond to the touch sensor if the arm was moving that fast.

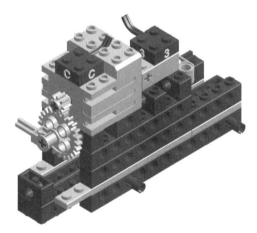

Figure 16-9:
Vending Machine Step 8

The base of Vending Machine is started in Step 9. The base has three functions: reading a card, supporting the RCX, and supporting the dispenser. By Step 11 the base is starting to take shape. The 2x2 bricks and 2x4 plates near the back of the base will eventually support the RCX. The inverted beam provides a smooth surface for the cards to slide in on, and the 1x16 beam near the front of the base will eventually support one side of the dispenser.

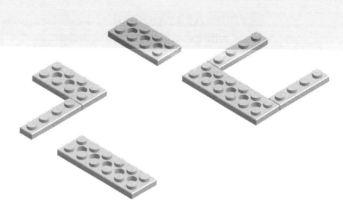

Figure 16-10:
Vending Machine Step 9

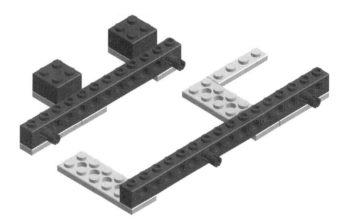

Figure 16-11:
Vending Machine Step 10

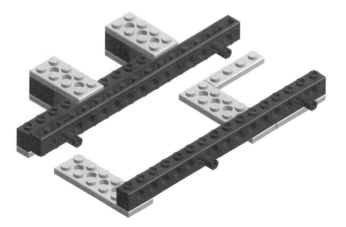

Figure 16-12:
Vending Machine Step 11

Steps 12 through 14 continue to develop the base. A light sensor will be used to detect the card's position near the back of the reader. In order to minimize interference from stray light, the rear portion of the reader is enclosed in bricks rather than beams. Note the two 1x2 beams near the back of the base. As usual, two plates separate these beams from the 1x16 beams below, thus providing the correct spacing for vertical bracing.

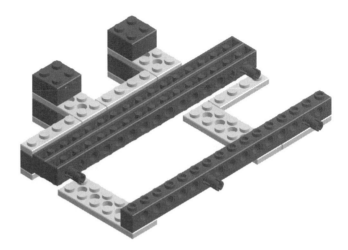

Figure 16-13:
Vending Machine Step 12

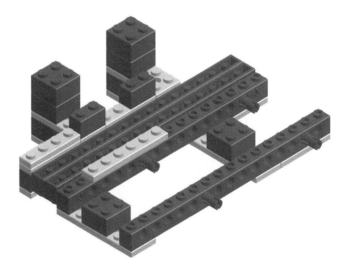

Figure 16-14:
Vending Machine Step 13

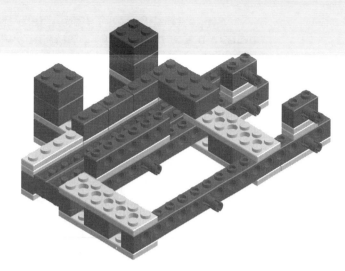

Figure 16-15:
Vending Machine Step 14

All of the card reader's electronics are added in Step 15. When a card is placed at the front of the reader, it will activate the touch sensor. The motor can then be used to pull the card in while the touch sensor counts how many 1x1 plates are on the card. This use of a tire is something new. In previous robots, tires were used as a way of moving the robot. Being made of rubber, they have much better grip than a smooth LEGO piece. In addition, their "sponginess" allows them to grip uneven surfaces (such as the top of a 2x10 plate). As a result, they are ideal for grabbing and pulling items through a machine. There are several different tires that will fit on the hub; be sure to use the smallest size. The light sensor detects when the card has been fully inserted, after which the motor can be reversed to eject the card.

Figure 16-16:
Vending Machine Step 15

Steps 16 through 18 complete the base. Note how the motor is held in place by the 2x4 plates above it. The 1x6 and 1x8 beams serve as two legs to hold the dispenser at the proper angle. As with Brick Sorter, the angles required do not quite work out to a natural right triangle, so one pair of legs is not quite perpendicular to the base.

Figure 16-17:
Vending Machine Step 16

Figure 16-18:
Vending Machine Step 17

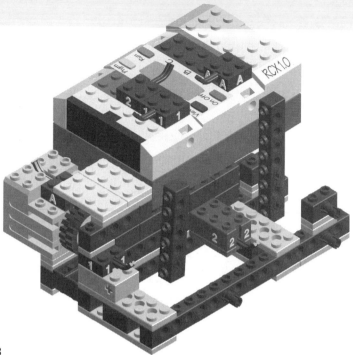

Figure 16-19:
Vending Machine Step 18

Vending Machine is completed in Step 19 by adding the dispenser and two more support legs. As usual, the orientation of motor wires is particularly important. Step 18 shows the ends of motor A's wire, while Step 19 shows the ends of motor C's wire. These orientations provide the most obvious path for the wire to reach its motor and they also ensure that forward rotation of the motors corresponds to what the programs expect. Specifically, the forward rotation of the reader motor should pull the card in and the forward rotation of the dispenser motor should turn the arm counterclockwise.

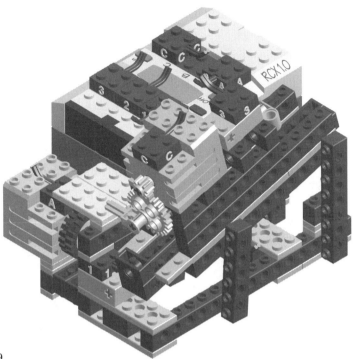

Figure 16-20:
Vending Machine Step 19

The dispenser was designed to accommodate Crispy M&Ms®. Other candies may require slight adjustments. Possible adjustments include the size and position of the rotating arm, the size of the opening at the bottom of the chute, the height of the guardrails, or even the size and angle of the chute itself. For example, adding a bushing on the end of the rotating arm makes it more suitable for working with smaller candies.

THEORY OF OPERATION

Card reading is triggered by the reader's touch sensor (which will become pressed if a card is placed at the opening of the reader). At this point, the program needs to turn on the reader's motor in order to pull the card into the reader. The light sensor is used to determine when the card is fully inserted, and the motor must be stopped at this point. While the card is being pulled in, the number of times the touch sensor is pressed must be counted since this determines how many candies should be dispensed.

Once the card has been read, candies can be dispensed by running the dispenser's motor. If set up properly, each rotation of the arm will result in a single candy being pushed out of the dispenser.

The exact number of rotations can be monitored by counting the number of times the dispenser's touch sensor is released. Waiting for the sensor to be released (rather than pressed) insures that the arm has pushed the candy out of the dispenser before we stop the motor.

Once the candy has been dispensed, the card may be ejected from the reader. This can be accomplished by running the reader motor in the reverse direction for a little while (one second is plenty of time). Just to be safe, the motor should continue running until the reader's touch sensor becomes released.

PROGRAMMING IN NQC

The NQC program can be broken into separate functions for the different phases of operation: reading the card, dispensing candy, and ejecting the card. Reading the card can be a little tricky because the program must monitor both the touch sensor and light sensor while the card is being inserted. A first attempt at programming this in NQC might look something like this:

```
#define CARD_COUNT    SENSOR_1
#define CARD_STOP     SENSOR_2
#define CARD_GRAB     OUT_A

#define THRESHOLD     40

int count;

void readCard()
{
    // wait for card to be inserted, then grab it
    until(CARD_COUNT == 1);
    OnFwd(CARD_GRAB);

    count = 0;
    while(CARD_STOP < THRESHOLD)
    {
        // count number of presses
        if (CARD_COUNT == 1)
            count++;
    }

    Off(CARD_GRAB);
}
```

Unfortunately, this doesn't quite work. The while loop may execute several times during the time it takes a single 1x1 plate to pass by the sensor. In order to count the actual presses we must somehow keep track of whether the button was previously pressed or not, and only count new presses. One such solution follows:

```
int lastValue, currentValue;
lastValue = 0;
while(CARD_STOP <= THRESHOLD)
{
    currentValue = CARD_COUNT;
    if (lastValue != currentValue)
    {
        if (currentValue == 1) count++;
        lastValue = currentValue;
    }
}
```

Two variables are introduced: lastvalue is used to remember the previous touch sensor value, and currentvalue is used to latch in the sensor's value. Setting lastvalue to 0 initially insures that the first 1x1 plate will be counted. In effect, a small state machine has been constructed inside of the while loop. A state machine can be described as a set of states and a set of transitions between states. In addition, one or more actions may be associated with a transition between states. In our example, there are 2 possible states (0 and 1) which are stored in lastvalue. A transition occurs when the current sensor reading is different from the state. If the transition is from state 0 to state 1, an action is performed (count is incremented). In either case, lastvalue is updated to reflect the new state. Although our example was rather simple, state machines provide a powerful way to describe complex systems and are used quite extensively in programming.

The RCX's multitasking capability can be exploited to solve this same problem another way:

```
void readCard()
{
    // wait for card to be inserted, then grab it
    until(CARD_COUNT == 1);
    OnFwd(CARD_GRAB);
```

```
        // start the card counting task
        start countTask;

        // continue grabbing until completely inserted
        until(CARD_STOP > THRESHOLD);
        Off(CARD_GRAB);

        // stop the counting task
        stop countTask;
    }

    task countTask()
    {
        count = 0;

        while(true)
        {
            until(CARD_COUNT == 1);
            count++;
            until(CARD_COUNT == 0);
        }
    }
```

In this case, a separate task waits for the sensor to read 1, then increments and waits for the sensor to return to 0. This task is explicitly started and stopped as needed by the readCard function. Since this task runs in parallel with the main task, the readCard function needs only to watch the light sensor and determine when to stop reading.

Both approaches have their advantages, although in such a simple example there isn't a compelling reason to prefer one to the other. The remainder of this chapter will continue to use the multi-task case since it is similar to how programs in RCX Code are constructed, and is slightly smaller (in RCX bytecodes).

Our second function must dispense the candy. As mentioned previously, the dispenser's touch sensor allows the RCX to monitor the number of revolutions of the rotating arm. As with the card reader, the program waits for the sensor to be pressed then released. The dispensing function is shown below:

```
#define CANDY_COUNT    SENSOR_3
#define CANDY_MOTOR    OUT_C
```

```
void dispenseCandy(const int & n)
{
    // dispense the candy
    On(CANDY_MOTOR);
    repeat (n)
    {
        until(CANDY_COUNT == 1);
        until(CANDY_COUNT == 0);
    }
    Off(CANDY_MOTOR);
}
```

Although previous programs have used functions, this is the first time a nonempty *argument* list is present. When a function is defined, one or more *formal arguments* may be declared. In this case, the formal argument n is declared. Each formal argument must have its type specified as well—in this case const int &. This rather strange type is just a terse way of saying that the value is passed by reference but never changed within the function.

When a function is called, it must be provided with the appropriate *actual arguments* that are substituted for the formal arguments during the execution of the function. Here are a few examples of calling the dispenseCandy function:

```
dispenseCandy(2);            // dispense 2 candies

int x;
dispenseCandy(x);            // dispense "x" candies

dispenseCandy(SENSOR_1);     // dispense based on sensor value
```

The details of substituting the actual arguments for the formal ones are called *argument passing semantics*, and depend on the declared types of the arguments. In fact, since all values in NQC are 16 bit integers, the only information truly specified in the type are the argument passing semantics. There are four possible types for arguments: int, int &, const int and const int &. Complete descriptions of the semantics of each of these types can be found in chapter 3. Additional information is also available in the *NQC Programmer's Guide*, which can be found on the CD-ROM for this book.

Type	Semantics	Comments
int	by value	uses up a variable, actual argument is not altered
int &	by reference	efficient, actual argument must be a variable
const int	n/a	efficient, actual argument must be constant
const int &	by reference	efficient, argument cannot be modified

Figure 16-1:
NQC Argument Types

Since dispenseCandy does not need to modify the argument, using const int & gives us flexibility in calling the function while still generating a very efficient code. Using a type of int would latch the actual argument into a temporary, which is often helpful when dynamic quantities are being used. However, in the case of dispenseCandy, the formal argument is only accessed a single time (setting the repeat count), thus latching has little advantage.

At this point argument passing may seem a bit overkill. After all, why not just let dispenseCandy directly use the global variable count? The answer is that using arguments can make a function much easier to reuse later on. Right now we want to dispense a number of candies based on a global variable count, but later on perhaps we'll want to use the function some other way. Perhaps the variable will be named candy, or perhaps we'll want to call it several times with a different variable each time. In fact, in the next chapter, we'll reuse dispenseCandy and call it with something that isn't even a variable.

Our third function is used to eject the card. It turns on the card reader's motor for a little while, then waits for the card to be removed before shutting off the motor.

```
void ejectCard()
{
    // eject card
    OnRev(CARD_GRAB);
    Wait(100);
    until(CARD_COUNT == 0);
    Off(CARD_GRAB);
}
```

All that remains is to add a suitable main task which calls readCard, dispenseCandy, **and** ejectCard **in order. The complete program is shown below:**

```
// vending.nqc
// program for the Vending Machine using card reader

#define CARD_COUNT    SENSOR_1
#define CARD_STOP     SENSOR_2
#define CARD_GRAB     OUT_A

#define CANDY_COUNT   SENSOR_3
#define CANDY_MOTOR   OUT_C

#define THRESHOLD 40

int count;

task main()
{
    // setup sensors
    SetSensor(CARD_STOP, SENSOR_LIGHT);
    SetSensor(CARD_COUNT, SENSOR_TOUCH);
    SetSensor(CANDY_COUNT, SENSOR_TOUCH);

    while(true)
    {
        readCard();
        dispenseCandy(count);
        ejectCard();
    }
}

void readCard()
{
    // wait for card to be inserted, then grab it
    until(CARD_COUNT == 1);
    OnFwd(CARD_GRAB);

    // start the card counting task
    start countTask;

    // continue grabbing until completely inserted
    until(CARD_STOP > THRESHOLD);
    Off(CARD_GRAB);
```

```
        // stop the counting task
        stop countTask;
    }

    task countTask()
    {
        count = 0;

        while(true)
        {
            until(CARD_COUNT == 1);
            count++;
            until(CARD_COUNT == 0);
        }
    }

    void dispenseCandy(const int & n)
    {
        // dispense the candy
        On(CANDY_MOTOR);
        repeat  (n)
        {
            until(CANDY_COUNT == 1);
            until(CANDY_COUNT == 0);
        }
        Off(CANDY_MOTOR);
    }

    void ejectCard()
    {
        // eject card
        OnRev(CARD_GRAB);
        Wait(100);
        until(CARD_COUNT == 0);
        Off(CARD_GRAB);
    }
```

The threshold for the light sensor, as usual, may require some adjustment. Pick a value midway between the sensor's reading when no card is inserted and when the card is directly in front of the sensor.

PROGRAMMING IN RCX CODE

The process of reading the card requires two stack controllers: one to wait for the initial press, and the other to wait for the light sensor to detect when the card has been fully inserted. Since RCX Code 1.0 has a limit of one stack controller per stack, we need to break this process up into two separate stacks (each in a *my command*). The first command, **waitcard**, waits until the reader's touch sensor is pressed, then resets the counter (which is used to record the number of bumps on the card). The second command, **grabcard**, activates the reader motor, waits for the light sensor to detect the card, then shuts off the motor. With the addition of a sensor watcher to increment the counter, our card reading process is complete. Note how the sensor watcher increments the counter when the sensor is released, rather than when it is pressed. This is because **waitcard** will be running in parallel with the sensor watcher, and if they are both trying to take action after the sensor is pressed, the counter may get cleared just before or just after the sensor watcher increments it. By making the sensor watcher respond to the release of the touch sensor, the operation becomes reliable.

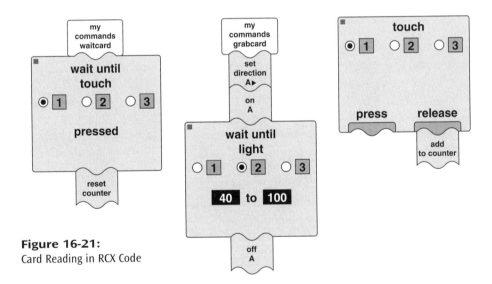

Figure 16-21:
Card Reading in RCX Code

Dispensing the candy is a little more difficult. Ideally we'd like to use the value of the counter as a repeat value, but RCX Code does not support this. If we were able to decrement the counter, then we could use a **repeat while** loop to implement this, but again we are thwarted by RCX Code's limitations. The solution is to change the meaning of the cards slightly. Instead of dispensing a number of candies equal to the value of the card, we'll turn things around and dispense a number equal to 5 minus the value of the card. For example, a card with value 4 would result in 1 candy, and a card with value 2 would result in 3 candies. Now we can use a stack controller to wait until the counter equals 5. The card reader will have already incremented the counter to a value between 1 and 4. A sensor watcher on the dispenser's touch sensor will continue to increment the counter each time a candy is dispensed.

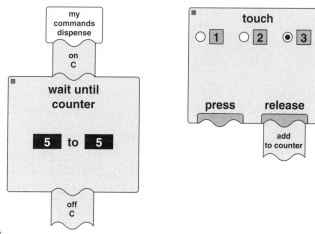

Figure 16-22:
Candy Dispensing in RCX Code

All that remains is ejecting the card and putting everything together inside a main loop. This code is shown below.

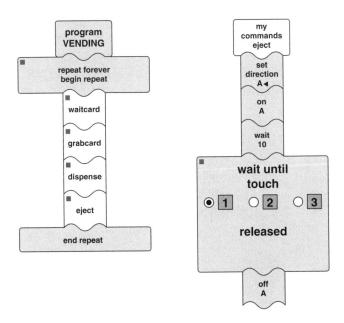

Figure 16-23:
Vending Machine
Program in RCX Code

TESTING

Vending Machine is now ready for use. Start with the arm pointing up and load some candy into the dispenser. Run the program and place a card at the front of the reader.

The reader motor should immediately start pulling in the card. If the motor pushes out the card, then the orientation of one of the wire's ends need to be changed. If the motor does not turn on at all, then verify that the card is constructed properly and activates the touch sensor. Another possibility is that the motor immediately ejects the card without first pulling it in. If this is the case, try raising the threshold for the light sensor. Assuming the card has been pulled in, the motor should stop when the card is fully inserted. If the motor is not stopping, then perhaps the light sensor threshold needs to be lowered.

Arranging things so that a single candy is dispensed with each rotation of the arm may require some adjustments depending on the specific details (weight, shape, and size) of the candy. If the candy is relatively heavy, then when the dispenser is full the weight of all of the candy may be sufficient to push the bottom one out of the dispenser. If this is the case, then the angle of the dispenser can be made less steep or the guardrails can be enlarged. If the candy is not sliding down into position fast enough, then the angle of the dispenser may need to be increased. A slightly smaller or larger arm may also be needed to ensure that a single candy is pushed out each time. The existing arm can be enlarged by adding one or more bushings. For a smaller arm, a 1x3 liftarm can be used.

If a single candy is being ejected with each rotation of the arm and Vending Machine is still dispensing the wrong number of candies, then the touch sensor isn't being activated reliably. The crossblock that presses against the touch sensor may require some slight adjustment so that it activates the touch sensor exactly once per rotation.

CONCLUSION

Vending Machine is my personal favorite of the robots in this book. It combines some familiar techniques (inverted beams, etc.) with a few new twists such as using a tire to "grab" an object and using a touch sensor to count rotations. Granted, the idea of dispensing M&Ms® in exchange for inserting a LEGO card is a bit contrived, but the robot is still fun to play with.

There are several obvious enhancements that can be built; perhaps a hopper to catch the candies as they are dispensed, or a large loading bin at the top of the chute for added capacity. The card reader itself works quite smoothly and could be easily integrated into other machines.

Vending Machine is used again in the next two chapters. In chapter 17, Vending Machine is reprogrammed to accept orders from either a computer or another RCX via infrared messages. In chapter 18, Vending Machine is programmed to maintain a record of all of its "transactions."

Chapter 17
Communication

The RCX is able to send and receive simple messages via infrared light. These messages can be exchanged with another RCX, or even a computer. This chapter presents two different projects that illustrate how to incorporate such communication into a robot.

In the first project, Vending Machine from chapter 16 will be reprogrammed to use messages instead of the card reader for determining when to dispense candy. Vending Machine can then be controlled remotely from a computer. Although this project can be programmed in RCX Code, the Robotics Invention System software itself cannot send a message from the computer to the RCX. Therefore some other means of sending messages to the RCX is required for testing the modified Vending Machine. There are several ways of doing this: using NQC to send the message from a computer, programming a second RCX to send it, or using the LEGO Remote Control (#9738) to send it directly.

In the second project, things get even more interesting as multiple RCXs communicate with one another. A robotic Delivery Truck will be created from Dumpbot, and Vending Machine will be rebuilt as a stand-alone Candy Dispenser. The Delivery Truck can accept an order from the user (via a touch sensor), drive to the dispenser, request the appropriate number of candies, then return and deliver its cargo. The programming required for this second project is too complex for RCX Code, so only NQC code is provided. In order to build both robots at the same time, a few extra pieces are required, most notably, a second RCX.

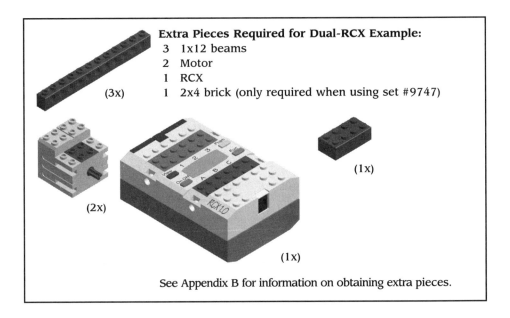

Extra Pieces Required for Dual-RCX Example:
3   1x12 beams
2   Motor
1   RCX
1   2x4 brick (only required when using set #9747)

(3x)

(2x)

(1x)

(1x)

See Appendix B for information on obtaining extra pieces.

This is a great project to work on with a friend who owns their own MINDSTORMS Robotics Invention System set. However, even if you are not able to complete the second project, reading through the material can still be instructive.

REPROGRAMMED VENDING MACHINE

Our first project is a modified Vending Machine that accepts orders via an infrared message instead of the card reader. If you still have Vending Machine built from the last chapter, you may use it as is. However, if you have not built Vending Machine, you may find it easier to build Candy Dispenser as described in the next section, rather than building the entire Vending Machine. Both Vending Machine and Candy Dispenser use the same dispensing motor and sensor arrangement, which is the important point from a programming perspective.

Before delving into the program itself, we need to understand how the RCX uses messages. Each message has a value between 0 and 255. The RCX is always listening for new messages and will remember the value of the most recent message it received (the value is 0 if no message has been received). This value can be used just like a sensor's value. For example, an RCX Code program can have a sensor watcher that responds to a specific message value, or an NQC program can use the value of the message in an if statement. It is also important for the RCX to be able to respond to another message once the current one has been handled. For this, commands to reset the message value back to 0 are provided (**reset message** in RCX Code, and `ClearMessage` in NQC).

We are now ready to make Vending Machine respond to messages. In RCX Code all we need is a sensor watcher for the message (called the "RCX" watcher). The watcher is set to respond to any message (value 1 to 255), and will then dispense a candy. Dispensing of the candy is done similarly to that with the original Vending Machine. A sensor watcher is used to increment the counter each time a candy has been dispensed. In this case, only a single candy needs to be dispensed, so the counter is first reset, then the motor is turned on and the program waits for the counter to hit 1. After dispensing the candy, the message is reset so that the program will be able respond to a new message. The complete program is shown below:

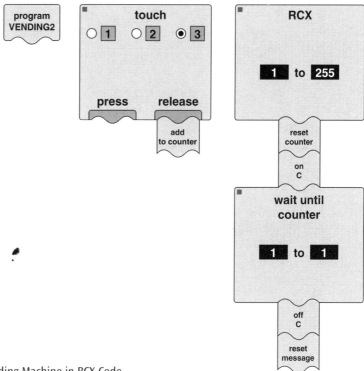

Figure 17-1:
Message-Based Vending Machine in RCX Code

Our NQC program can be a bit more sophisticated than its RCX Code counterpart. Instead of dispensing a single candy for each message, it can use the value of the message itself as the number of candies to dispense. The complete program is shown below:

```
// vending2.nqc
// program for message based Vending Machine

#define CANDY_COUNT    SENSOR_3
#define CANDY_MOTOR    OUT_C

task main()
{
    // setup sensors
    SetSensor(CANDY_COUNT, SENSOR_TOUCH);

    while(true)
    {
        // wait for message
        until(Message() != 0);

        // dispense candy
        dispenseCandy(Message());

        // get ready for next message
        ClearMessage();
    }
}

void dispenseCandy(const int &n)
{
    // dispsense the candy
    On(CANDY_MOTOR);
    repeat (n)
    {
        until(CANDY_COUNT == 1);
        until(CANDY_COUNT == 0);
    }
    Off(CANDY_MOTOR);
}
```

Message always returns the value of the most recently received message, and it remains 0 until a message is received. The program then calls dispenseCandy (copied directly from the original

Vending Machine program), using Message as the argument. Finally, ClearMessage is called to set the message value back to 0 and get ready for another message.

Regardless of whether NQC or RCX Code was used to program Vending Machine, we still need one more thing before we can test the program: a way to send a message to the RCX. The specifics of doing this depend greatly on the programming tools you are using. For example, if you are using the command line version of NQC, simply type the following command line substituting the desired message value for *value*:

```
nqc -msg value
```

If you are using RcxCC, select **Send Messages** from the **Tools** menu. This will bring up a window from which you can send messages. Click on the appropriate number button, or use the arrows to enter a number at the bottom of the window and press the **Send** button.

In MacNQC, select **Test Panel** from the **RCX** menu to show the test panel window. In the **Message** section of the window enter the appropriate number and press the **Send** button.

If you are using the Robotics Invention System software, then things are a bit more difficult. RCX Code allows you to write RCX programs that send and receive messages, but sending a message from the computer to the RCX is not supported. If you happen to have an extra RCX lying about, the following program will turn it into a message sender:

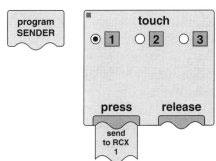

Figure 17-2:
Sending a Message in RCX Code

Hook a touch sensor up to sensor port 1, run the program, and press the touch sensor to send a message. A second option is to use the LEGO Remote Control, which can be used to send three different messages. The buttons labeled "1," "2," and "3" on the remote control correspond to sending messages of values 1, 2, and 3, respectively.

A final option is to use one of the NQC programming tools (NQC, RcxCC, or MacNQC) to send the message. Inside the RCX, both NQC and RCX Code use the same messages, so programs written in NQC can communicate with programs written in RCX Code and vice versa. Furthermore, the NQC tools can be used to send a message from the computer to an RCX Code program. For this you'll need to install the appropriate tools (see chapter 1) and follow the instructions given above for sending messages.

CANDY DISPENSER

Our second communication example uses two robots: Candy Dispenser and Delivery Truck. Candy Dispenser is really just a stripped-down version of Vending Machine with the dispenser raised high enough to extend over Delivery Truck. The dispenser portion of Vending Machine, shown in Step 1, can be used with only one minor modification: both of the short wires used in Vending Machine should be replaced with long wires.

Figure 17-3:
Dispenser Step 1

A new base for the dispenser must then be built as shown in Steps 2 through 5. The bottom 2x8 plate gives the base a slight "T" shape, which increases its side-to-side stability. The only other subtle detail is the 2x4 plate on top of the 1x2 beams. It provides little structural value, but it does make assembly a little easier by holding the 1x2 beams in place when it is time to add vertical beams. Sometimes good design is just as much about the assembly process as the final product.

Figure 17-4:
Dispenser Step 2

Figure 17-5:
Dispenser Step 3

Figure 17-6:
Dispenser Step 4

Figure 17-7:
Dispenser Step 5

The dispenser is attached to the base using a pair of 1x10 beams and a pair of 1x12 beams. As in our previous examples, one pair of beams (1x12s) is kept perpendicular while the other pair (1x10s) is allowed to form a slightly different angle. The RCX sits off to the side and is connected to the dispenser by two long wires. This allows the RCX to be positioned so that it can receive messages, and the dispenser positioned to drop candy into Delivery Truck.

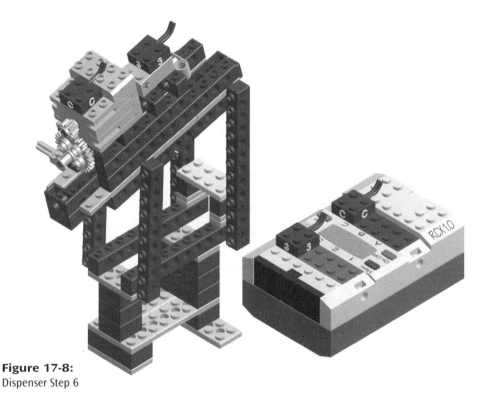

Figure 17-8:
Dispenser Step 6

Candy Dispenser uses the vending2.nqc program from the previous example with a slight twist. In our scenario Delivery Truck will send a message to Candy Dispenser, then wait for the candy before turning around and returning home. How long should Delivery Truck wait? The solution is to have Candy Dispenser tell Delivery Truck when it is done by sending a message back. The complete program follows:

```
// dispense.nqc
// program for Candy Dispenser
```

```
#define CANDY_COUNT    SENSOR_3
#define CANDY_MOTOR    OUT_C

task main()
{
    // setup sensors
    SetSensor(CANDY_COUNT, SENSOR_TOUCH);

    while(true)
    {
        // wait for message
        until(Message() != 0);

        // dispense candy
        dispenseCandy(Message());

        // get ready for next message
        ClearMessage();

        // tell Delivery Truck we are done
        SendMessage(1);
    }
}

void dispenseCandy(const int &n)
{
    // dispense the candy
    On(CANDY_MOTOR);
    repeat (n)
    {
        until(CANDY_COUNT == 1);
        until(CANDY_COUNT == 0);
    }
    Off(CANDY_MOTOR);
}
```

DELIVERY TRUCK

Delivery Truck requires a trivial extension to Dumpbot: the only thing added is a touch sensor connected to sensor port 3. This touch sensor is how orders are given to Delivery Truck.

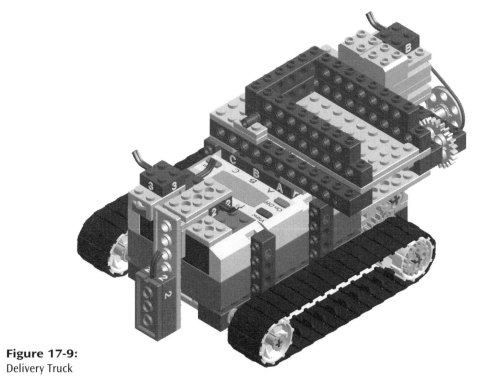

Figure 17-9:
Delivery Truck

Delivery Truck's mission is to accept an order from the user, follow a line until it reaches Candy Dispenser, request the desired number of candies, turn around and return home, and dump its cargo. The functions for following a line, turning around, and dumping the cargo bin were previously developed as part of the dumpbot2.nqc program. The new requirements are to accept an order and communicate with Candy Dispenser.

A touch sensor is used to input orders to Delivery Truck—the "order" is simply the number of times the button is pressed. Counting button presses is easy, but how will the program know when to stop counting and start driving? We will use a *timeout*, which means that if the program does not detect a new button press after a certain amount of time, it will assume that the order is complete.

Counting the presses is done in its own task:

```
#define BUTTON SENSOR_3

int order;
```

```
task counting_task()
{
    // count the number of button presses
    while(true)
    {
        until(BUTTON == 1);
        ClearTimer(0);
        PlaySound(SOUND_CLICK);
        order++;
        until(BUTTON == 0);
    }
}
```

The code waits for the touch sensor to be pressed, then it clears a timer (used later on for the timeout), plays a sound (as feedback to the user), increments the order count, and waits for the touch sensor to be released. Once started, this task will constantly monitor the touch sensor, keeping track of the total number of presses.

A separate function takes care of starting and stopping the task and manages the timeout:

```
#define TIMEOUT 20

void get_order()
{
    // start with a clean order
    order = 0;

    // start counting
    start counting_task;

    // wait until at least 1 press and then a timeout
    until(order == 1);
    until(Timer(0) >= TIMEOUT);

    // no more counting
    stop counting_task;
}
```

Note how the function first waits for at least one button press, then waits for the timer to exceed the timeout (in this case 2 seconds). The complete program (including functions from dumpbot2.nqc) looks like this:

```
// delivery.nqc
// program for the Delivery Truck

// linebot sensor, motors, and constants
#define EYE              SENSOR_2
#define LEFT             OUT_A
#define RIGHT            OUT_C
#define LINE_THRESHOLD              51
#define STOPPER_THRESHOLD           42
#define TURN_SPEED                  3
#define INITIAL_TIME  4

// dumper motor and constants
#define DUMPER          OUT_B
#define DUMP_TIME       70
#define UNDUMP_TIME     100

// sensor and constant for taking an order
#define BUTTON SENSOR_3
#define TIMEOUT 20
#define REVERSE_TIME  80

void dump()
{
    // dump the cargo bin
    OnFwd(DUMPER);
    Wait(DUMP_TIME);
    Rev(DUMPER);
    Wait(UNDUMP_TIME);
    Off(DUMPER);
}

int direction, time, eye, ok_to_stop;

sub follow_line()
{
```

```
// initialize the variables
direction = 1;
time = INITIAL_TIME;
ok_to_stop = 0;

// start driving
OnFwd(LEFT+RIGHT);

while(true)
{
    // read the sensor value
    eye = EYE;

    if (eye < LINE_THRESHOLD)
    {
        // we are either over the line or a stopper
        if (eye > STOPPER_THRESHOLD)
            ok_to_stop = 1;
        else if (ok_to_stop == 1)
        {
            // found a stopper
            Off(RIGHT+LEFT);
            return;
        }
    }
    else
    {
        // need to find the line again
        ClearTimer(0);
        if (direction == 1)
        {
            SetPower(RIGHT+LEFT, TURN_SPEED);
            Rev(LEFT);
        }
        else
        {
            SetPower(RIGHT+LEFT, TURN_SPEED);
            Rev(RIGHT);
        }

        while(true)
        {
```

```
                            // have we found the line?
                            if (EYE < LINE_THRESHOLD)
                            {
                                time = INITIAL_TIME;
                                break;
                            }

                            if (Timer(0) > time)
                            {
                                // try the other direction
                                direction *= -1;
                                time *= 2;
                                break;
                            }
                        }

                        SetPower(LEFT+RIGHT, OUT_FULL);
                        Fwd(RIGHT+LEFT);
                    }
                }
}

void turn_around()
{
    // start turning
    OnRev(LEFT);
    OnFwd(RIGHT);

    // wait until not over line
    until(EYE >= LINE_THRESHOLD);

    // wait until over line again
    until(EYE < LINE_THRESHOLD);
    Off(LEFT+RIGHT);
}

int order;

task counting_task()
{
    // count the number of button presses
    while(true)
    {
```

```
            until(BUTTON == 1);
            ClearTimer(0);
            PlaySound(SOUND_CLICK);
            order++;
            until(BUTTON == 0);
        }
}

void get_order()
{
    // start with a clean order
    order = 0;

    // start counting
    start counting_task;

    // wait until at least 1 press and then a timeout
    until(order == 1);
    until(Timer(0) >= TIMEOUT);

    // no more counting
    stop counting_task;
}

task main()
{
    SetSensor(EYE, SENSOR_LIGHT);
    SetSensor(BUTTON, SENSOR_TOUCH);

    get_order();

    follow_line();

    // send request, wait for response
    ClearMessage();
    SendMessage(order);
    until(Message() != 0);

    // turn-around
    Rev(LEFT+RIGHT);
    OnFor(LEFT+RIGHT, REVERSE_TIME);
    turn_around();
```

```
        // return to starting place and dump
        follow_line();
        dump();

        PlaySound(SOUND_UP);
    }
```

The `main` task is straightforward. The order is accepted, then the line is followed until a stopper is encountered. The program then sends the request to Candy Dispenser and waits for a reply. Note how `ClearMessage` is called before sending the message to Candy Dispenser. The call to `ClearMessage` is required so that an old message doesn't cause the program to think a reply has been sent before Candy Dispenser is really done. Before turning around, Delivery Truck backs up a little (0.80 second). This gets Delivery Truck clear of Candy Dispenser before it starts turning around. Finally, the program follows the line back to the other stopper, dumps the cargo, and plays a special sound to signify that it is done.

TESTING

Testing Delivery Truck requires some sort of line to follow, with stoppers at either end. As with Dumpbot (and Linebot before it), the line should be ¾" thick and colored gray or green on a white surface. Curves should have a 6" or larger radius, and stoppers (which are black) should be at least ¾" long. As usual, the thresholds for the light sensor will need to be adjusted. If Delivery Truck is having trouble following the line, refer to chapter 8.

Candy Dispenser needs to be positioned at one end of the line. For best results, a short straight section of line (at least 12") should precede this end. This will ensure that Delivery Truck always approaches Candy Dispenser straight on. Run Delivery Truck and make sure it stops at the end of the line. Now set Candy Dispenser's RCX so that it is a few inches away and facing Delivery Truck. The dispenser itself should be positioned as close as possible to Delivery Truck, so that the candies will fall into the cargo bin. I created a simple test mat on a piece of 22" x 28" poster board as shown below:

A small dish was also placed on the mat to catch the candies as they were dumped from Delivery Truck's cargo bin. It may take a few rounds of adjustment to get Candy Dispenser placed just right. If it is too close, Delivery Truck may bump into it. If it is too far away, some candy may miss the dish.

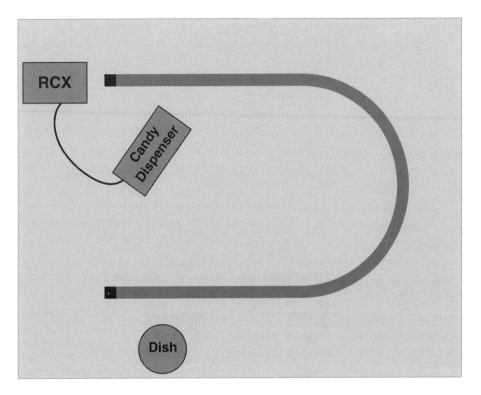

Figure 17-10:
Test Mat for Delivery Truck

UNLIMITED HORIZONS

Communication is perhaps the most powerful feature of the RCX. It allows the RCX to accept commands from a computer, or multiple RCXs to cooperate to accomplish a joint mission. In either case, messages provide a way to overcome some of the RCX's inherent limitations. For example, the RCX normally can only control 3 motors and use 3 sensors. If a certain mission requires more motors and/or sensors, multiple RCXs can be used, coordinating their activities via messages.

Communication with a computer can be used to overcome limitations in the RCX's own computational power. Data can be sent from the RCX to the computer for processing, which can then relay commands back down to the RCX via messages. At one extreme, the computer can even take over "primitive" control of the robot— turning on and off each motor individually. A more sophisticated approach would be to have the robot performing semi-autonomously: an order could be sent to the RCX, which would then execute a complete sequence of actions and report back to the host computer.

Chapter 18
Using the Datalog

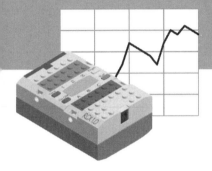

The *datalog* allows the RCX to record data that can later be uploaded to a computer. Using the datalog, it is possible to turn the RCX into a data collection device. Existing robots can be adapted to log information about their operation, or new devices can be built specifically for detecting and recording information.

In this chapter, three uses for the datalog will be explored. In the first example, the message-based Vending Machine from chapter 17 (or Candy Dispenser) will be augmented to keep track of each candy request. This transaction list can then be uploaded to a computer for analysis (perhaps to figure out how much more candy to buy). This example illustrates how the datalog can be used to supplement an existing robot's functionality.

In the second example, the RCX will keep track of how often and for how long a television is turned on. In this case, the RCX is being used merely as a data collection device. The third example takes the data collection a step further. In it, a temperature sensor is used to watch the temperature of water as it is being frozen. This example illustrates how the datalog can be used to assist in performing experiments. The temperature sensor is not part of the MINDSTORMS Robotics Invention System, and must be purchased separately. The temperature sensor isn't nearly as versatile as the other sensors, so you should only get one if you are really interested in building some temperature based robots. See Appendix B for more information on how to obtain a temperature sensor.

Unfortunately, RCX Code does not provide access to the datalog. Therefore, only NQC versions of the sample programs will be provided.

DATALOG BASICS

The datalog must be initialized before any data can be added to it. Initialization consists of telling the RCX how many entries the datalog should hold. The datalog is shared across all programs and retained even when the RCX is turned off. The datalog requires 3 bytes of memory per entry, and this memory comes from the same pool used to store programs (about 6,000 bytes total). If several large programs are loaded into the RCX, then there may not be enough room for the desired datalog. Conversely, a large datalog may prevent certain programs from being loaded. The largest possible datalog would hold over 2000 entries, but such a large datalog would leave almost no room for programs.

The NQC function to initialize the datalog is CreateDatalog, which takes a single argument, the desired size (number of entries). The RCX may only hold a single datalog at any given time, so each call to CreateDatalog erases the previous datalog and then creates a new datalog of the specified size. The following statement initializes the datalog to hold 100 entries:

```
CreateDatalog(100);   // create datalog to hold 100 entries
```

A size of 0 indicates that no datalog should be created. This is a good way to get rid of the previous datalog when memory is getting tight.

```
CreateDatalog(0);     // eliminate datalog
```

Initialization of the datalog reserves enough memory to hold the specified number of entries. The datalog itself is empty after initialization. Data can be added using the AddToDatalog command. A sensor reading is the most obvious thing to add, but any other value (such as a timer or variable) may be added as well.

```
AddToDatalog(SENSOR_1);    // add the value of sensor 1
AddToDatalog(Timer(0));    // add the current timer value
AddToDatalog(Watch());     // add the system watch
AddToDatalog(7);           // add a constant
```

Data can continue to be added until the datalog is full, after which any additional calls to AddToDatalog will be ignored. When data is added, the datalog records the source of the data as well as the data value itself. There are 4 data sources that the datalog understands directly: variables, timers, sensors, and the system watch. Any other type of value (such as a constant) is first moved into a variable, then the variable is added to the datalog. The NQC compiler takes care of this detail for you, but you need to be aware of it when you look at the datalog later, since logging a constant (or other nonsupported type) will appear in the datalog as a variable.

The RCX's display contains a datalog indicator which shows if a datalog is currently in use, and if so how much data has been added. The indicator is a small circle divided into 4 sections which appears at the right side of the display. When no datalog exists, all 4 sections are turned off. When the datalog is in use, each section represents approximately one quarter of the datalog. As each quarter of the datalog is filled, the appropriate section is shown. In addition, the quarter where the next entry will be added is indicated by a flashing section. For example, if a datalog for 100 entries was created and 60 entries have been added, 2 sections will be solid, and a third section will be flashing. Once the datalog is completely full, all 4 sections will be solid.

The datalog can be uploaded to a computer at any time, even if it is not yet full. The upload is typically initiated from the computer, and the specifics depend on the programming tool being used. When using the command line version of NQC, the following command will upload the current datalog and print it on screen:

```
nqc -datalog
```

Each line will contain a single datalog entry value, formatted as a signed decimal integer. In order to view the sources for the entries as well as their values, use the following command:

```
nqc -datalog_full
```

To upload the datalog from RcxCC, select **Datalog** from the **Tools** menu and press the **Upload Datalog** button. The datalog will be uploaded and the sources and values of the entries will appear within the datalog window.

In MacNQC, the datalog can be uploaded by selecting **Upload Datalog** from the **RCX** menu. The datalog will be uploaded and displayed in its own window. If the **Verbose Datalog** preference is set, then both the sources and values will be displayed, otherwise only the values will be shown. The preferences can be edited by selecting **Preferences** under the **Edit** menu.

LOGGING TRANSACTIONS

The first datalog example adds a transaction record to the message based Vending Machine from the last chapter. However, since the dispensing mechanisms and programming are so similar, the same changes could also be made to either the previous chapter's Candy Dispenser, or even the original Vending Machine (chapter 16).

Two features need to be added to the program: initialization of the datalog, and adding entries to the datalog. Initialization of the datalog can be very simple—adding a single line to the main task should do the trick:

```
CreateDatalog(100);
```

However, this has two disadvantages. The first is that we will have no way of knowing if the call is successful. The RCX does not provide a way for a program to detect if the datalog initialization was successful, but the user can always check by looking for the datalog indicator on the LCD display. Unfortunately, if a previous (and smaller) datalog already exists, it is possible that initialization could fail but the display would still indicate a datalog is present (the old one). This would mislead the user into thinking the proper size datalog was created when in fact it wasn't. We can work around this by first erasing the old one, then creating the new one:

```
CreateDatalog(0);
CreateDatalog(100);
```

This code will still fail in the same situations as the first one (namely, not enough memory). The difference is that when the code fails, the previous datalog will have already been deleted so the display will indicate no datalog present. Therefore the user can tell immediately from the LCD display if the desired datalog has been created or not.

The second disadvantage of such an initialization is that a new datalog is created every time the program is run. What if we want the log to span several different runs of the program? One way to do this is to have a completely separate program initialize the datalog. The user could then run this other program whenever a new datalog was desired. A better solution is to let the program figure out if the user wants a new datalog. Consider the following function:

```
#define CANDY_COUNT    SENSOR_3
#define DATALOG_SIZE   100

void init_datalog()
{
    // if not pressed, don't init datalog
    if (CANDY_COUNT == 0) return;

    CreateDatalog(0);
    CreateDatalog(DATALOG_SIZE);
    PlaySound(SOUND_FAST_UP);

    // wait for button to be released
    until(CANDY_COUNT==0);
}
```

The function first checks to see if the CANDY_COUNT sensor (the touch sensor on the dispenser portion of Vending Machine) is released. Under normal operation, Vending Machine will always stop with this sensor released, so the only way it would be pressed is if the user is intentionally holding it in. If it is released, then the function simply returns, doing nothing. However, if the sensor is pressed, then a new datalog is created, a sound is played (to provide some feedback to the user), and the function waits for the sensor to be released. Assuming that init_datalog is called fairly early from within the main task, the user can cause a new datalog to be created by pressing and holding the touch sensor before running the program. Once the sound is played, it is safe to release the touch sensor.

So much for initialization of the datalog. We also need to add data to it. We'll keep it simple and only add the number of candies dispensed for each request. A good place to do this is within the dispenseCandy function:

```
void dispenseCandy(const int &n)
{
    // log the request
    AddToDatalog(n);
```

```
    // dispsense the candy
    On(CANDY_MOTOR);
    repeat (n)
    {
        until(CANDY_COUNT == 1);
        until(CANDY_COUNT == 0);
    }
    Off(CANDY_MOTOR);
}
```

This has one potential problem. In certain situations it is possible that the value of n will change between the AddToDatalog command and the repeat statement. For example, in the case where dispenseCandy is called using Message as the argument (as in chapter 17), it is possible that a new message would be received at just the wrong moment. This problem is very unlikely, and could probably be ignored, but I like to be on the safe side. The best way to combat this problem is to latch the value of n. We could do this explicitly by creating another variable and copying n to the new variable. Another option is to change the type of n so that dispenseCandy is called by value (which means the compiler will create a variable and do the copying for us).

```
void dispenseCandy(int n)
{
    // log the request
    AddToDatalog(n);

    // dispsense the candy
    On(CANDY_MOTOR);
    repeat (n)
    {
        until(CANDY_COUNT == 1);
        until(CANDY_COUNT == 0);
    }
    Off(CANDY_MOTOR);
}
```

The complete program for a message-based Vending Machine with datalog support is shown below. The choice of using a datalog size of 100 is arbitrary. It is large enough to do plenty of experimenting, but small enough not to cause any severe memory shortages. Feel free to use another size if it suits your needs better.

```
// vending3.nqc
// program for message based Vending Machine with data logging

#define CANDY_COUNT    SENSOR_3
#define CANDY_MOTOR    OUT_C

#define DATALOG_SIZE   100

task main()
{
    // setup sensors
    SetSensor(CANDY_COUNT, SENSOR_TOUCH);

    init_datalog();

    while(true)
    {
        // wait for message
        until(Message() != 0);

        // dispense candy
        dispenseCandy(Message());

        // get ready for next message
        ClearMessage();
    }
}

void init_datalog()
{
    // if not pressed, don't init datalog
    if (CANDY_COUNT == 0) return;

    CreateDatalog(0);
    CreateDatalog(DATALOG_SIZE);
    PlaySound(SOUND_FAST_UP);

    // wait for button to be released
    until(CANDY_COUNT==0);
}

void dispenseCandy(int n)
{
    // log the request
    AddToDatalog(n);
```

```
// dispsense the candy
On(CANDY_MOTOR);
repeat (n)
{
    until(CANDY_COUNT == 1);
    until(CANDY_COUNT == 0);
}
Off(CANDY_MOTOR);
}
```

Be sure to press and hold the dispenser's touch sensor (the one attached to input 3) when running the program for the first time. If all goes well, you should hear the sound being played and the datalog will be initialized. If the sound was not played, then the program is not trying to initialize the datalog. Check the LCD display. It should show a single flashing datalog section. If the sound played, but the LCD display does not show any datalog indicators, then there isn't enough memory for the datalog. Try using a smaller value for DATALOG_SIZE in the program and/or freeing up memory by removing unwanted programs.

The most thorough way to clear out memory is to erase everything, then reload whichever program (or programs) are needed. To clear memory using the command line version of NQC, use the following command:

```
nqc -clear
```

In RcxCC, the same function can be accomplished by selecting **Clear Memory** from the **Tools** menu. In MacNQC, select **Clear Memory** from the **RCX** menu.

Beware that clearing memory using any of these methods will erase all programs in the RCX as well as any previous datalog.

Assuming that the datalog was initialized properly, use Vending Machine (or Candy Dispenser) a few times, then upload the datalog. You should see a list of all the requests processed since the datalog was created. For example, a transcript using the command line version of NQC is shown below:

```
> nqc -msg 2
> nqc -msg 4
> nqc -msg 1
> nqc -datalog
2
4
```

```
1
> nqc -msg 3
> nqc -datalog
2
4
1
3
```

KEEPING TRACK OF THE TELEVISION

Our second example shows how the datalog enables the RCX to be used as a data collection device. The first thing to determine is what data should be collected. It just so happens that my television has a small red LED that lights up whenever the television is turned on. By placing a light sensor up against this LED, the RCX will be able to use the light sensor to determine when the television is on. We could record the light sensor's value periodically (e.g., once a minute), but this would lead to a large amount of data—much of which has little value. A more efficient way of keeping track of the television is to just record those instances when the television's state changes: going from "off" to "on", and vice versa. For each of these events, the time at which the even occurred will be recorded in the datalog. There are two different ways of measuring time in the RCX: the timers and the system watch. The timers provide resolution of 10 ticks per second, and have a maximum value of 54.6 minutes. The watch has a resolution of 1 minute and a maximum value of 23 hours and 59 minutes. The watch is a good match for our application, so we'll use it.

```
// tvlog.nqc
// program that logs television usage

// some constants
#define TV_SENSOR     SENSOR_1
#define THRESHOLD     40
#define DATALOG_SIZE  100

task main()
{
    // setup sensors
    SetSensor(TV_SENSOR, SENSOR_LIGHT);

    // create the datalog
```

```
CreateDatalog(0);
CreateDatalog(DATALOG_SIZE);

while(true)
{
    // wait for TV to be turned on and log it...
    until(TV_SENSOR > THRESHOLD);
    AddToDatalog(Watch());
    PlaySound(SOUND_CLICK);

    // now wait for TV to be turned off...
    until(TV_SENSOR < THRESHOLD);
    AddToDatalog(Watch());
    PlaySound(SOUND_CLICK);
}
}
```

For each transition from "off" to "on," or "on" to "off," the program will add the current time (according to the system watch) to the datalog. The threshold for the light sensor can be adjusted by changing the value of THRESHOLD. Similarly, the capacity of the datalog can be adjusted by changing DATALOG_SIZE.

The system watch resets to 00.00 (0 hours and 0 minutes) each time the RCX is turned on. The watch may be set to some other, perhaps more meaningful, value. When using the command line version of NQC, use the following command, substituting the desired time for *hhmm*.

```
nqc -watch hhmm
```

For example, to set the watch to 1:30, you would use the following command:

```
nqc -watch 130
```

Another option is to automatically set the watch to the current time (at least the current time according to the computer).

```
nqc -watch now
```

To set the watch when using RcxCC, select **Diagnostics** from the **Tools** menu. You can either enter a new value for the watch manually, or press the **Set Current** button to set it to the current time. In MacNQC, the watch can be set from the test panel (select **Test Panel** from the **RCX** menu). Enter the desired value under **Watch**, then press **Set** to send the new value to the RCX.

In order to conserve batteries, the RCX will automatically shut off after a certain period of inactivity (known as the *power down time*). Curiously enough, running a program is not enough "activity" to keep the RCX from turning itself off, therefore the power down time will need to be changed from its initial setting of 15 minutes if any significant amount of logging is to be done. To change the power down time using the command line version of NQC, use the following command, substituting the desired number of minutes:

```
nqc -sleep minutes
```

Setting the power down time to 0 disables the feature.

```
nqc -sleep 0
```

If you are using RcxCC, the power down time can be changed by selecting **Diagnostics** from the **Tools** menu and entering the appropriate time in the dialog box. In MacNQC, the power down can be set from the test panel (select **Test Panel** from the **RCX** menu). Enter the desired value under **Sleep** and press the **Set** button to send the new value to the RCX. In all cases, a time of 0 minutes indicates that the RCX should never turn itself off.

Speaking of batteries, the RCX consumes batteries the fastest when it has to power motors. In this experiment, no motors are involved so power consumption is relatively small. Even so, running an RCX program for 20 or 30 hours is likely to drain batteries completely. If you plan on running a data logging program for extended periods of time, consider using an AC adapter (if available for your RCX). LEGO makes an appropriate adapter (see Appendix A for more information).

After setting up the RCX such that the light sensor was directly in front of my television's power indicator, I let the program run for a while. Here's a sample datalog:

```
1110
1140
1260
1320
```

The watch records time in minutes using a 24-hour clock. For example, 1:00 AM would be 60, 2:00 AM would be 120, and 1:00 PM would be 780. The above log entries correspond to the following times:

6:30 PM
7:00 PM
9:00 PM
10:00 PM

This indicates that the television was turned on twice: the first time at 6:30 PM for 30 minutes, and the second time for 60 minutes at 9:00 PM.

This program could also be used in other situations; perhaps monitoring how much time a room's lights are turned on, or how often a refrigerator is opened.

WATCHING WATER FREEZE

The next example makes use of the temperature sensor to observe the process of water freezing. Before digging into the specifics of the experiment, it is important to understand some of the limitations of this particular sensor. The temperature sensor reacts much slower than the other sensors. This is because the sensor is really measuring the temperature of the sensor itself, which is not necessarily the same temperature as whatever environment it is in. It always takes some amount of time for the sensor to reach the same temperature as its environment. Consider what happens if the sensor is suddenly put in the freezer. Although the temperature of the freezer is quite cold, the sensor will initially still be at room temperature. Over time, the sensor will cool off until it reaches the ambient temperature of the freezer itself.

The time it takes for the sensor to equalize to a new environment depends on many factors including the change in temperature and the properties of the material the sensor is in contact with. The graph below shows temperature readings taken at 1 minute intervals as the sensor was moved from room temperature into a refrigerator. In the first case, the sensor was simply placed inside the refrigerator and was thus in contact only with air. In the second case, the sensor was immersed in a small dish of water that had already been in the refrigerator for some period of time.

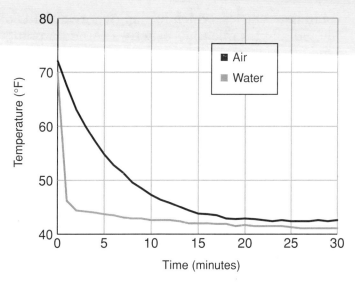

Figure 18-1:
Cooling Effects of Air and Water

Both cases reached different final temperatures, but this is not very significant. The reason is that only one touch sensor was used and the experiments were thus done at slightly different times. The temperature inside a refrigerator is not constant; it continually moves up and down a few degrees as it gradually warms up, then turns on the compressor and cools down. Therefore, the refrigerator was not at the same exact temperature for both tests.

The significant point is that water caused a much faster change in temperature. This is because water conducts heat much better than air. That is the reason that swimming in 70-degree water feels much colder than walking around in a 70-degree room. The water does a much better job of conducting your body heat away, thus making you feel colder.

The lesson here is that it can take quite a while, upwards of 15 minutes in this example, for the temperature sensor to settle in on close to the right value after a drastic temperature change. Good conductors, such as water, can lower this time to a couple of minutes, but this is still a far cry from the nearly instantaneous reactions of the touch, light, and rotation sensors.

In order to observe what happens as water freezes, we need a program that will collect temperature readings regularly over a large period of time. I chose to read the sensor every 15 seconds over a 6-hour period. By clearing out all other programs the resulting

1440 data entries, which use up over two thirds of the RCX's memory, should all fit. Collecting exactly 1440 data points can be a bit tricky. One attempt might be:

```
repeat(1440)
{
    // collect a data point
}
```

The problem with this is that repeat only works with numbers up to 255. Anything larger is an error that will be caught by the compiler. You can fool the compiler like this:

```
int x = 1440;
repeat(x)
{
    // collect a data point
}
```

But the RCX still won't repeat 1440 times. This is because the RCX only supports repeats up to 255, and will ignore the "overflow" when a larger value is supplied. You just can't use repeat loops for iterations of more than 255.

> When large (greater than 255) repeat counts are used, the actual number of repeats is a consequence of how the RCX stores numbers. Most values in the RCX are stored as 2 *bytes*. However, the repeat count is only stored as a single byte, so the second byte is discarded. The resulting count is equal to the remainder of the initial value divided by 256. For example, a repeat count of 257 becomes a repeat of 1. The repeat count of 1440 in the example becomes a repeat of 160.

What about nesting the loops?

```
repeat(6)
{
    repeat(240)
    {
        // collect a data point
    }
}
```

This would appear to make sense. We still have a total of 1440 data points, but never a repeat over 255. The problem is that the RCX just can't handle it—repeat loops cannot be nested at all. We need to abandon the idea of using a repeat loop.

The solution is to use a variable to keep track of the iterations and manually check it ourselves. The complete program is shown below:

```
// templog.nqc
// log temperature every 15 seconds for 6 hours

#define SAMPLE_TIME        1500   // 15 seconds
#define DATALOG_SIZE       1440   // 6 hours of data

task main()
{
    SetSensor(SENSOR_1, SENSOR_FAHRENHEIT);
    SelectDisplay(1);

    CreateDatalog(0);
    CreateDatalog(DATALOG_SIZE);

    int i = 0;
    while(i < DATALOG_SIZE)
    {
        AddToDatalog(SENSOR_1);
        Wait(SAMPLE_TIME);
        i++;
    }
}
```

Strictly speaking, the samples won't be exactly 15 seconds apart. Each iteration of the loop will take slightly longer than 15 seconds due to the time spent adding to the datalog and the over-head of the loop itself. This variation in timing is known as drift. In this case, the amount of drift is nearly insignificant compared to the time between samples (a few milliseconds every 15 seconds), so we can safely ignore it.

We're now ready to perform the experiment. Put some fresh batteries in the RCX (or use an AC adapter), get a small dish of water ready, and immerse the tip of the temperature sensor in the water. I used one quarter cup of distilled water and heated it up to about 100 degrees first. Tap water at room temperature (or any other temperature for that matter) could also be used. I placed the water (and temperature sensor) inside the freezer and left the RCX outside, then let the program run. The display will show the current temperature reading, which can tell you at a glance how things are going inside the freezer.

> Disclaimer: water and electronics generally don't mix. Ice can be even worse. I arranged things such that only the tip of the sensor got wet and the rest of it remained dry. I haven't noticed any ill effects yet, but you can never be sure. If you are concerned about potential damage to the sensor, then consider "sealing" it in a lightweight plastic bag. The plastic is an insulator and will obscure the readings a bit, but the general experiment can still be performed.

Six hours later, the datalog was ready to be uploaded to a computer. The first thing that you will notice about the datalog is that it contains rather large numbers.

```
995
977
955
937
917
901
...
```

This is because the RCX always uses integers. Since the temperature sensor measures with 0.1 degree of precision, the RCX internally works with the temperature multiplied by 10. Therefore, 99.5 degrees is represented as 995, hence the absurdly high datalog numbers.

The other problem with the datalog is that 1440 data points is a lot of data. Graphs with that many data points can be unwieldy. Thinning out the data to a sample every 5 minutes and dividing the temperatures by 10 we see the following:

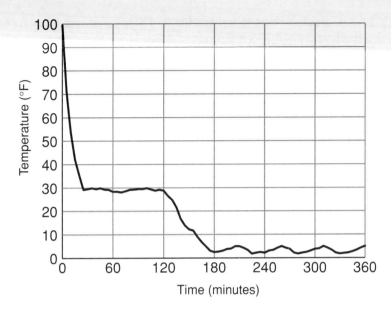

Figure 18-2:
From Water to Ice

Several things are evident from this graph. During approximately the first 25 minutes the water was rapidly being cooled down to roughly 30 degrees. At this point something interesting happened. Rather than continuing to cool, the water remained around 30 degrees for nearly 100 minutes. This was the period of time that the water was gradually turning to ice. In general, a mixture of water and ice will remain at the freezing point until all of the water has been frozen. This property of freezing lets us "see" just how long it took for the water to freeze.

> There are examples of water and ice systems that have varying temperatures, but the rule of thumb that the mixture remains at the freezing point holds for our simple ice making experiment.

Notice that it took about 4 times as long to freeze the water as it did to cool it 70 degrees. This is because *phase changes* such as going from water to ice (or water to steam) involve significantly more energy than just raising or lowering temperature.

Another interesting point is that the graph shows the freezing point at 30 degrees rather than 32 degrees. Does my freezer make ice at a different temperature than everywhere else in the world? No. This is because the temperature sensor is really not accurate to more than a few degrees. It is quite reliable in terms of relative temperature (hence the shape of the graph), but cannot be used to accurately establish an absolute temperature.

> If you duplicate this experiment yourself, it is very possible that your water will freeze at a temperature lower than 32 degrees. Salts and other impurities that are present in water will lower its freezing point. Using distilled water, however, ensures a minimum of impurities.

After the water turned completely to ice, the temperature of the ice continued to drop until it reached about 2 degrees, then oscillated between 2 and 5 degrees. This oscillation gives us some insight into how the freezer itself is working. The compressor (which cools the freezer) turns on at 5 degrees, then turns off at 2 degrees. Remember the discussion on feedback in chapter 8? The three degree difference between the on and off temperatures provides enough hysteresis so that the compressor turns on about every 50 minutes.

If we "zoom-in" to the period of time just before ice started forming we can observe another interesting phenomenon. The graph in Figure 18-3 shows all of the samples (every 15 seconds) between 15 and 35 minutes into the experiment.

Notice how the temperature of the water dips several degrees too low, then snaps up to the freezing point. During this time the water was still in liquid form at a temperature below the freezing point. This is known as *supercooling*. Ice crystals don't spontaneously form as soon as water reaches the freezing point. Some other factor, such as an impurity in the water, or a rough surface, is the impetus for starting the crystallization. When fewer impurities are present, it may take the water molecules a little longer to start forming, thus allowing for supercooling. Once the first crystal forms, additional ones form quite rapidly around the first one. The result of this is that once the ice starts forming, the temperature of the mix-

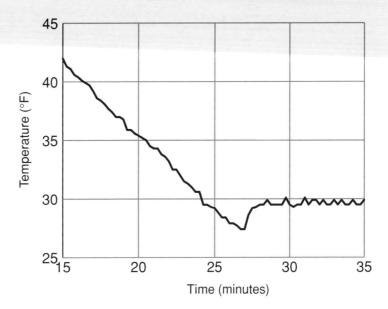

Figure 18-3:
Supercooling

ture will quickly rise back up to the freezing point. In fact, the temperature probably rose faster than the graph indicated because the temperature sensor itself is somewhat slow to respond to changes.

CONCLUSION

Three different applications for the datalog were presented. In the first example, the datalog was used to record transactions in a robot. A variation of this approach can be useful in debugging a robot's program. At various points in the program, information can be added to the datalog. Later on, the datalog can then serve as a record of what bits of the program were being executed and in what order.

The second example showed how events could be logged over a long period of time. This technique can also be adapted for a variety of purposes. The third example probably sounded more like a physics lesson than a discussion of robotics, but it is another good illustration of how the RCX can be used for data collection purposes.

Many robots won't have any use for the datalog. However, for those few situations where data needs to be collected, the datalog is an indispensable feature of the RCX.

Chapter 19
Roboarm

Roboarm is mechanically the most complex robot presented in this book. Its three motors allow it to spin around looking for objects, grab and release them, and lift them up and down. Two touch sensors are used to determine the position of the arm, and a light sensor allows Roboarm to "see" objects to pick up.

Owners of #9747 will need to provide a few extra bricks.

1.5

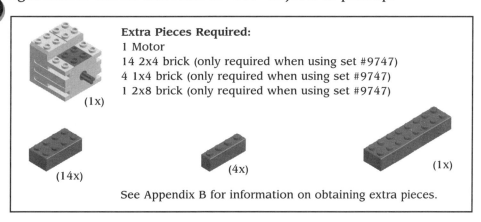

Extra Pieces Required:
1 Motor
14 2x4 brick (only required when using set #9747)
4 1x4 brick (only required when using set #9747)
1 2x8 brick (only required when using set #9747)

(1x)

(14x)   (4x)   (1x)

See Appendix B for information on obtaining extra pieces.

Roboarm also uses more pieces than any of the other robots. Therefore it is especially important to use the correct pieces in each step; otherwise you may run out of a critical piece later on. Pay particular attention to using regular yellow 2x4 plates (rather than the gray ones with holes) wherever appropriate. Also note the difference between 1x2 bricks (no hole) and 1x2 beams (with a hole).

THE BASE

Roboarm needs a stable base underneath it. In order for Roboarm to be able to turn freely, the base must have a smooth top. By contrast, the bottom of the base needs to grip well, so that it will remain firmly planted as the arm itself rotates. In order to meet these requirements, the base is built in a slightly unorthodox manner. The sides of bricks will become the top of the base, and wheels will be placed face down on the bottom of the base to provide a grip. In addition, beams are used within the base itself for increased strength.

The base was designed such that it could be built from either set #9719 or #9747 with a minimum of extra parts. Set #9719 includes a number of 2x8 bricks which may be substituted for the combinations of bricks in the bottom and top layers of the base. Using these longer bricks will result in a more sturdy base. Similarly, if you are using set #9747 and have some other LEGO bricks on hand, feel free to substitute larger bricks for combinations of smaller ones wherever feasible.

Steps 1 through 3 show the initial construction of the base. Note how two 1x2 beams with friction pegs are embedded near the bottom of the base. In later steps, another pair of beams and pegs will also be embedded near the top. Long beams will then be attached between these pegs as a way of reinforcing the structure. The center bricks are narrower than the ones on the side; this will result in a recessed area within the center of the base.

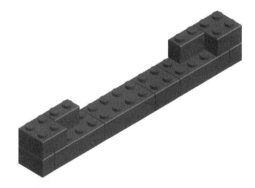

Figure 19-1:
Roboarm Base Step 1

Figure 19-2:
Roboarm Base Step 2

Figure 19-3:

Roboarm Base Step 3

So far, the base has been symmetrical, but in Step 4 a small gap is created on one side. This will eventually be used to allow the robot to detect its current position with respect to the base. This gap is extended into the next layer in Step 5.

Figure 19-4:
Roboarm Base Step 4

Figure 19-5:
Roboarm Base Step 5

The upper half of the base is like the bottom half, with the exception that there is no gap and a layer of plates has been added. The reason for the plates is that 1x12 beams will be used to brace the structure. The distance between the end holes of a 1x12 beam is 10 LEGO units. Since a brick is 1.2 LEGO units high, it will take a stack of 8.33 bricks to fill this same distance—in other words we need 8 bricks and 1 plate.

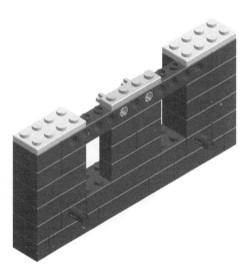

Figure 19-6:
Roboarm Base Step 6

Figure 19-7:
Roboarm Base Step 7

Figure 19-8:
Roboarm Base Step 8

The 1x12 beams added in Step 9 brace the structure and pro-vide a way to attach the wheels shown in Step 10.

Figure 19-9:
Roboarm Base Step 9

Figure 19-10:
Roboarm Base Step 10

For the final step, the base is placed facedown. The #6 axle will become the axis of rotation for Roboarm. Note how the 40-tooth gear is locked into position by the pair of axle pegs. During assembly it may be easier to remove these pegs, insert them into the gear, and attach the combined gear and pegs to the base. Roboarm's body will later use an 8 tooth gear meshed with the base's 40-tooth gear to rotate itself. One other detail is the small alcove that resulted from the gaps introduced back in Steps 4 and 5. This alcove is used by Roboarm's body to determine a home position while spinning.

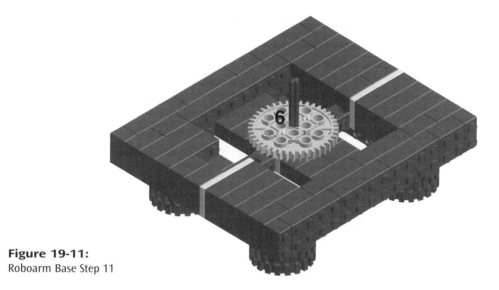

Figure 19-11:
Roboarm Base Step 11

The base is now complete and can be set aside for later.

THE ARM

As its name implies, Roboarm needs an arm. The arm has three basic design requirements. First of all, it must be capable of being raised and lowered. Second, it must be able to grasp and release objects. Third, it needs to be able to "see" objects within its grasp.

Raising and lowering an arm can be a lot of work. In fact, the longer the arm is, the more torque is required to lift it. This is because the arm is actually a lever, and the torque required to lift a given weight is proportional to the weight's distance from the axis of rotation. In other words, a 2-foot arm will require twice as much torque as a 1-foot arm (assuming all other factors are the same).

Generating a lot of torque is not too difficult. The output of a standard motor can be put through a gear train of some sort to reduce speed and increase torque. Transferring the resultant torque to the arm is another matter. Torque is the product of force and distance from the axis. Thus, the amount of torque generated by a given force increases the farther away from the axis the force is applied. This is the reason that long wrenches and screwdrivers with fat handles are better at loosening tight nuts and bolts. The simplest way to lift the arm is to attach a gear to one end and then rotate the gear. There is a practical limitation, however, to the total amount of force that can be applied to the teeth of a standard LEGO gear. For this reason, it would be wholly inadequate to use a small 8-tooth gear. The large 40-tooth gear is a much better choice because it has a much larger diameter, and thus can generate much more torque given the same force on its teeth.

Steps 1 through 4 show how the rear portion of the arm is constructed, including how a pair of 40 tooth gears are anchored to the beams that form the arm's structure. The #6 axle serves as the axis of rotation for the arm, and the #4 axle is used to lock the gears to the beams. Although a single 40-tooth gear would probably be sufficient in most cases, sharing the load between a pair of gears provides a more robust design capable of lifting larger weights. Two other important details are the 1x2 brick and half peg added to the near side of the arm in Step 4. These will be used later on by Roboarm to detect the position of the arm.

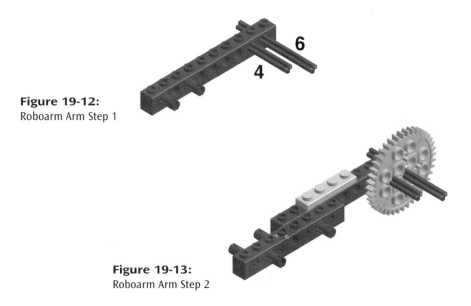

Figure 19-12:
Roboarm Arm Step 1

Figure 19-13:
Roboarm Arm Step 2

Figure 19-14:
Roboarm Arm Step 3

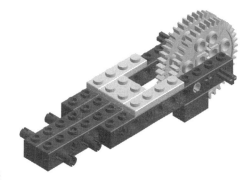

Figure 19-15:
Roboarm Arm Step 4

The arm must be able to grab objects, and this function requires its own motor. Although the grabbers will be at the far end of the arm, the motor itself will be placed closer to the axis used to lift the arm. Once again, we need to think in terms of torque. The motor's weight is actually quite significant relative to other LEGO pieces. Placing this motor farther out on the arm increases the torque required to lift it. Keeping the motor close to the axis minimizes its effect on the arm, and it is an easy enough matter to connect it to the grabbers by using a #8 axle (shown here) and a #12 axle (introduced later). The motor is coupled to this axle using a white rubber belt and 2 friction pulleys. This does not affect the speed or torque (apart from the effect of friction), but it does allow the motor to slip once the grabbers have closed around an object. Attach one end of a long wire to the motor as shown. This wire will be attached to the RCX later.

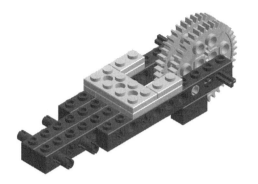

Figure 19-16:
Roboarm Arm Step 5

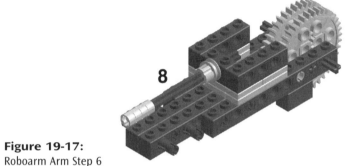

Figure 19-17:
Roboarm Arm Step 6

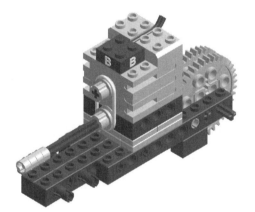

Figure 19-18:
Roboarm Arm Step 7

    The remaining length of arm is constructed in Steps 8 and 9, along with the grabbers. One grabber uses a pair of crossblocks at its end, but the other uses a pair of cross catches. The difference is not significant other than the fact that a total of four parts is required and the Robotics Invention System only includes two each of the crossblock and the cross catch.

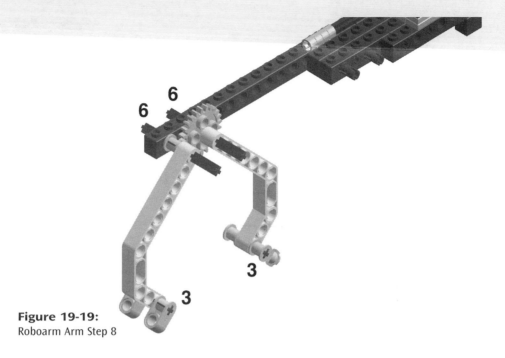

Figure 19-19:
Roboarm Arm Step 8

The pair of 16-tooth gears added in Step 9 keep the two sides of the grabber aligned with one another. It is important to position the grabbers properly before locking both gears into place. A good way to do this is to start by placing only one of the gears. Close the grabbers so that they meet at a point directly below the axle for the rear grabber (closer to the motor). Then, slip the second gear into place, adjusting the grabbers slightly in order to get the two gears to mesh. Turning the 24-tooth gear should now result in opening and closing both grabbers, and they should be roughly centered. Note that it is impossible to get the grabbers to be perfectly symmetrical. This is due to the fact that the gears have an even number of teeth, and, in order to mesh, the axles must be at slightly different angles.

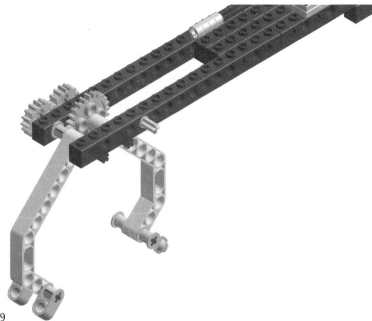

Figure 19-20:
Roboarm Arm Step 9

To give the grabbers a better grip, tires are stretched around the crossblocks and cross catches at the ends of the grabbers. Be sure to use the smallest tires; otherwise the grabbers will not be able to close properly. The grabbers are powered by a worm gear meshed with the 24-tooth gear. This results in a 24:1 gear reduction, which provides the grabbers with a reasonable amount of force before the rubber belt will slip. The worm gear also operates as a one-way transfer of power. The motor can easily spin the worm gear, which will result in a rotation of the 24-tooth gear. The reverse, however, is impossible. No amount of force applied to the 24-tooth gear will cause the worm gear to spin. This means that the motor can be used to open and close the grabbers, but that, once closed, the motor is no longer required for the grabber to maintain its grip.

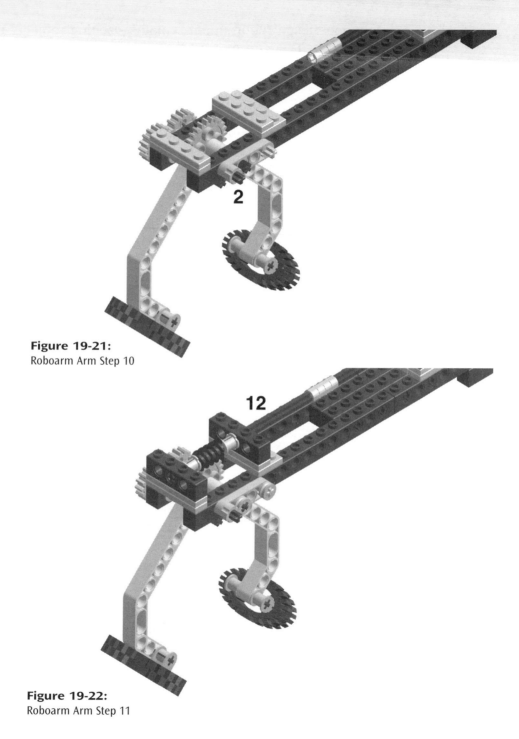

Figure 19-21:
Roboarm Arm Step 10

Figure 19-22:
Roboarm Arm Step 11

Only one design requirement remains: "seeing" objects within the arm's grasp. This can be accomplished by mounting the light sensor on the side of the arm such that it is aimed roughly between the grabbers, as shown in Steps 12 through 14. The light sensor is attached to a pair of 1x2 cross beams (the green ones), which in turn are connected to a 1x3 liftarm. Ideally, a #3 axle would be used to join the cross beams and the liftarm; however all of the available #3 axles are needed elsewhere, so a #2 axle is used instead. This means that it will only be partially inserted into the first cross beam shown in Step 12.

Figure 19-23:
Roboarm Arm Step 12

Figure 19-24:
Roboarm Arm Step 13

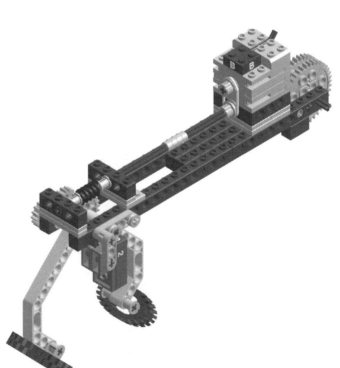

Figure 19-25:
Roboarm Arm Step 14

The arm is now complete and can be set aside until we are ready to attach it to the rest of Roboarm.

THE BODY

Roboarm's body is where most of the action takes place. The motors for spinning and lifting are with the body, along with suitable gear reduction. A touch sensor is mounted flush with the bottom of the body to detect the home position when rotating on the base. A second touch sensor is used to determine when the arm has been lowered to a level position. Construction begins with the relatively straightforward design of a frame for the rest of the body.

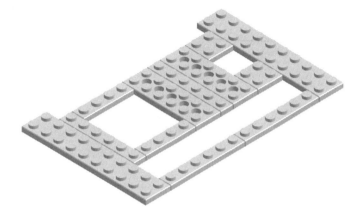

Figure 19-26:
Roboarm Body Step 1

Figure 19-27:
Roboarm Body Step 2

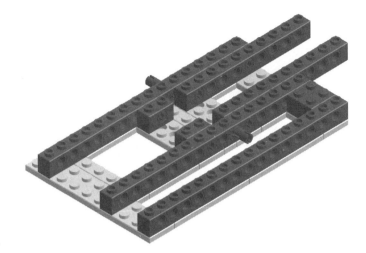

Figure 19-28:
Roboarm Body Step 3

The motor to be used for rotating the arm is added in Step 4. Eventually, the body will sit on top of the base, and it will rotate itself via an 8-tooth gear that meshes with the 40-tooth gear embedded in the base. The 8-tooth gear must therefore be below the body itself. In a later step, the #6 axle added in Step 4 (hereafter called the *spinner axle*) will be pushed through to extend below the frame, and the 8-tooth gear will be added. For now, however, it is convenient for the frame to sit level on a table, so everything will be built as if the axle did not need to extend below the frame. It is also worth noting that when the body is mounted on the base, the

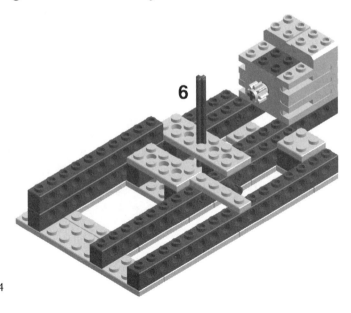

Figure 19-29:
Roboarm Body Step 4

main axle of the base will extend up through the center hole of the 2x4 plate just added.

The next few steps build up several features in parallel. A number of gears are used to transfer power from the motor to a 40-tooth gear on the spinner axle. First the 8-tooth gear on the motor meshes with a 24-tooth gear (3:1 gear reduction). This gear is on the same axle as a crown gear (24-teeth) that meshes with the 40-tooth gear (3:5 reduction). This results in a 5:1 overall gear reduction between the motor and the spinner axle.

The touch sensor for detecting the rotational position of the body is mounted flush with the bottom of the frame. Its position is such, that once the body is placed on the base, the touch sensor will be pressed except when it is directly over the alcove in the base. This allows Roboarm to detect its home position. A second motor for lifting the arm is also added. For now, this motor simply turns a worm gear.

Attach one end of a short wire to each of the motors and the touch sensor as shown. Later on, these three wires will be attached to the RCX.

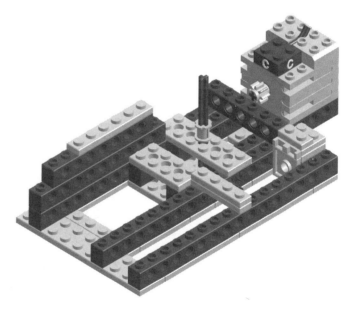

Figure 19-30:
Roboarm Body Step 5

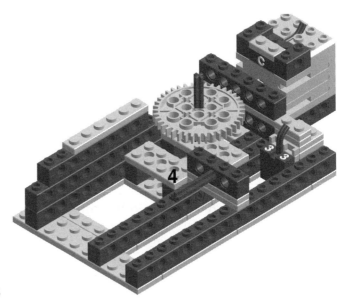

Figure 19-31:
Roboarm Body Step 6

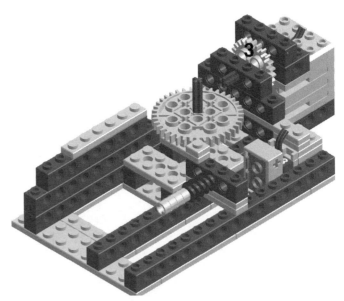

Figure 19-32:
Roboarm Body Step 7

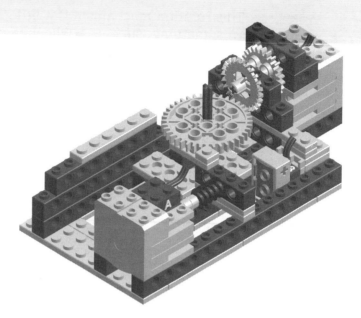

Figure 19-33:
Roboarm Body Step 8

The next step is to place the body on top of the base. First, the spinner axle must be pushed downward enough that an 8-tooth gear can be added below the body. The body is now ready to be placed on the base. Carefully align the body so that the axle from the base fits through the vacant hole in the body's 2x4 plate. The 8-tooth gear underneath the body must mesh with the base's 40-tooth gear. This 5:1 gear reduction combines with the 5:1 reduction between the motor and spinner axle for a total reduction of 25:1. Getting everything aligned may require the body to be turned slightly as it is lowered onto the base. If completed properly, the bottom of the body should be level and flush with the top of the base.

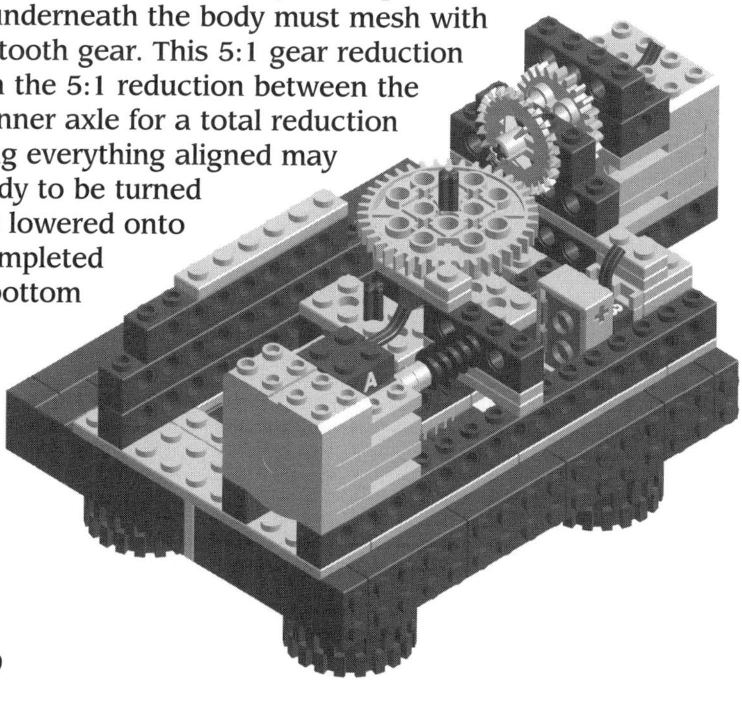

Figure 19-34:
Roboarm Body Step 9

This rotation mechanism is a little unusual. Previously, in order to turn something (such as a wheel), a series of gears was used to transfer power from the motor to the item to be turned. In this case, the motor powers the 8-tooth gear beneath the body, which is meshed to the 40-tooth gear in the base. If the base gear were free to spin, activating the motor would cause the base gear to turn. In this case, however, the base gear is not free to spin (it is anchored to the base); thus the entire base must rotate relative to the body. This can happen in one of 3 ways: the base can spin, the body can spin or some combination of both can spin. From a design perspective, we need to make sure that the body spins and the base remains motionless. This is why rubber tires were used on the bottom of the base—they ensure that the friction between the base and the table is great enough to keep the base firmly in position.

It can be instructive to manually turn the motor (or at least the 24-tooth gear it meshes with) and observe how the overall mechanism works.

A more traditional design could have used a motor in the base to turn a gear that was attached to the body, but this would cause an eventual wiring problem. Most of the motors and sensors are required to be part of the body and/or arm, which means they will be rotating with respect to the base. The RCX must be part of either the base (nonrotating) or the body (rotating). If one motor were contained in the base, then either RCX location would require at least one wire to go from the rotating portion to the nonrotating portion. To prevent the wire from becoming extremely twisted, software would need to ensure that the arm never completed too many revolutions in one direction before reversing direction. Since this is a somewhat annoying restriction to place on the operation of the arm, a design that allowed all of the motors to be on the rotating portion was used instead.

We are now ready to attach the arm to the body. Steps 10, 11, and 12 show how the arm is mounted to a simple frame. A pair of 8-tooth gears mesh with the arm's 40-tooth gears (5:1 reduction).

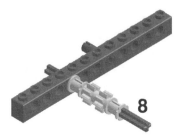

Figure 19-35:
Roboarm Body Step 10

Figure 19-36:
Roboarm Body Step 11

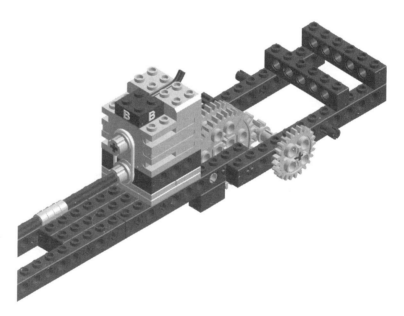

Figure 19-37:
Roboarm Body Step 12

The frame is then mounted to the body (Step 13) and secured in place with two 1x6 beams used for vertical bracing. The worm gear in the body now meshes with a 24-tooth gear (24:1 reduction), which combines with the previous 5:1 reduction for a total of

120:1 between the motor and the arm itself. This is more than enough torque to raise and lower the arm smoothly. Once again the one-way property of a worm gear is exploited; in this case it means that the angle of the arm is maintained even when the lifting motor is turned off.

A touch sensor and wire are also added on top of the lifting motor. This sensor is used to determine when the arm is level. The half peg and 1x2 brick previously added to the arm ensure that the sensor remains pressed whenever the arm is in a lifted position. Only when it is nearly level will the touch sensor be released.

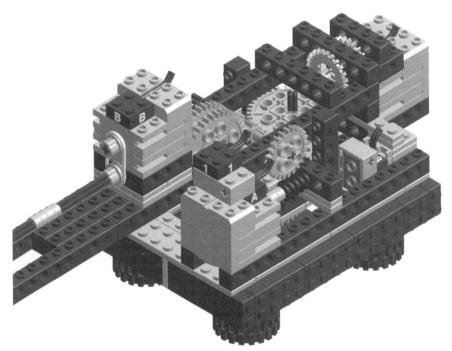

Figure 19-38:
Roboarm Body Step 13

The final step is to attach the RCX and wires. Position the RCX so that it is centered above the 1x6 and 1x4 beams near the back of the body. The wire on the light sensor is not long enough to reach to the RCX, so it must be extended by connecting it to a long wire.

Pay careful attention to the orientation of motor wires. On the RCX, all of the motor wires are attached such that the wire points to the front of the RCX (toward the LCD display). For sensor wires, orientation is not important. The sensor and motor assignments are as follows:

Input	Sensor	Wire
1	lifting touch sensor	short wire
2	light sensor	extend with long wire
3	rotation touch sensor	short wire

Output	Motor	Wire
A	lifting motor	short wire
B	grabber motor	long wire
C	rotation motor	short wire

In order to prevent the long wires from becoming tangled during operation of the arm, it is a good idea to coil up the excess length and secure it with a rubber band. The construction of Roboarm is now complete.

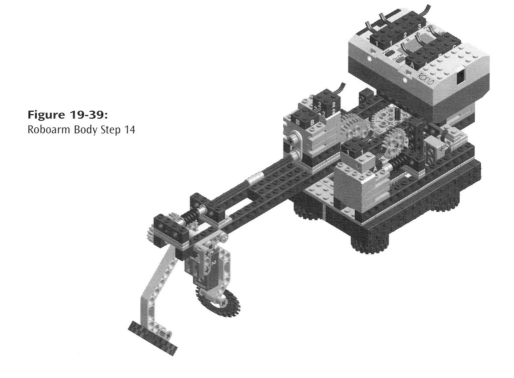

Figure 19-39:
Roboarm Body Step 14

BASIC OPERATIONS

Just as Roboarm is constructed from several simpler subassemblies, its programs are constructed from several basic operations, or programming "building blocks." Each operation performs a specific action such as closing the grabbers or lifting the arm. There are a total of five operations: **home**, **up**, **down**, **grab**, and **release**. **Home** spins the arm until the touch sensor indicates that it is in the home position. **Up** raises the arm a fixed amount by running the lifting motor for a short period of time. **Down** lowers the arm by running the lifting motor until the other touch sensor indicates the arm is approximately level. **Grab** and **release** close and open the grabbers by running a motor for a fixed amount of time. The time for grabbing is slightly longer than releasing to insure that the grabbers will always close firmly. The RCX Code stacks for these operations are shown below:

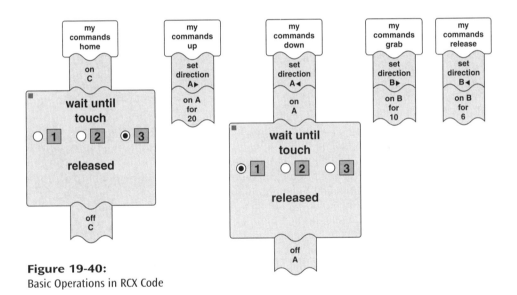

Figure 19-40:
Basic Operations in RCX Code

When using NQC, these basic operations will be placed in their own file (called "roboarm.nqh"), which will later be *included* by the actual programs. Keeping commonly used operations in their own separate file allows multiple programs to use them without having to write the same function repeatedly. It is a long-standing tradition in C programs to use the extension ".h" for included files (as opposed to ".c" for actual program files). We will emulate this tradition by using the ".nqh" extension for included files (instead of the more typical ".nqc" for programs).

```
// roboarm.nqh
// basic operations for Roboarm

// motors and sensors
#define LIFTER      OUT_A
#define GRABBER     OUT_B
#define SPINNER     OUT_C
#define LEVEL   SENSOR_1
#define EYE     SENSOR_2
#define HOME    SENSOR_3

// various timing parameters
#define UP_TIME       200
#define DOWN_DELAY    10
#define GRAB_TIME     100
#define RELEASE_TIME  60

void up()
{
    // raise arm
    Fwd(LIFTER);
    OnFor(LIFTER,UP_TIME);
}

void down()
{
    // lower arm until level
    OnRev(LIFTER);
    until(LEVEL==0);
    Wait(DOWN_DELAY);
    Off(LIFTER);
}

void grab()
{
    // close the grabbers
    Fwd(GRABBER);
    OnFor(GRABBER, GRAB_TIME);
```

```
}

void release()
{
    // open the grabbers
    Rev(GRABBER);
    OnFor(GRABBER, RELEASE_TIME);
}

void home()
{
    // spin until home
    On(SPINNER);
    until(HOME==0);
    Off(SPINNER);
}
```

The RCX Code and the NQC versions of the commands are similar. The only difference is that the NQC version of down leaves the motor running for a short time (DOWN_DELAY) after the sensor condition is met. This is because the touch sensor gets triggered slightly before the arm is level, and the NQC program reacts a little quicker than the RCX Code version. Thus the extra delay gets the arm closer to level.

If the arm raises up but never moves down, perhaps it is missing either the 1x2 brick or the half peg shown in Step 4 of the arm's construction (Figure 19-15).

STARTING POSITION

It is a bit easier to program the arm if it is already in a known starting position. A good choice for this position is to have the arm level and in its home position with the grabbers open. It is possible to manually adjust Roboarm by turning the worm gears, but it is more interesting to write a program to do this for us. First we consider the height of the arm. If the touch sensor for the arm height is pressed, then the arm may simply be lowered using the down operation. However, if the touch sensor is not pressed, the arm could be too low. In this case, we want to raise it first, then lower it to the level position. In the case of the grabbers, using the grab then release operations will generally result in their being open the proper amount regardless of their initial position. Rotating the arm into home position is the easiest of all; after all that is the entire purpose of the home operation. The following NQC program puts everything together in the proper order. Note how the `#include` statement is used to make everything previously defined in "roboarm.nqh" available to the program itself.

```
// roboarm1.nqc
// program put Roboarm in "starting position"

// include the basic operations
#include "roboarm.nqh"

task main()
{
    // configure the sensors
    SetSensor(LEVEL, SENSOR_TOUCH);
    SetSensor(EYE, SENSOR_LIGHT);
    SetSensor(HOME, SENSOR_TOUCH);

    // adjust arm height
    if (LEVEL==0)
        up();
    down();

    // close and open grabbers
    grab();
    release();

    // spin if not already home...
    home();
}
```

RCX Code no longer has this limitation. **1.5**

A similar program can be written in RCX Code, but, as usual, the stack limitations get in the way. Specifically, there's no convenient way to perform the conditional test before calling the **up** action. One alternative would be to use a **check and choose** for this, but then the remaining actions would have to appear in both the **true** and the **false** stacks. Instead, a new *my command* called **testup** is created to perform the test and optionally raise the arm. Note that **testup** cannot call **up**; thus it must reproduce the functionality of **up** block by block. The program itself is shown below. Note that it requires the **down**, **grab**, **release**, and **home** commands defined in the previous section. When using RCX Code 1.5, the **testup** *my command* can call up directly rather than reproducing its functionality block by block.

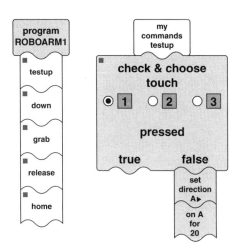

Figure 19-41:
Starting Position in RCX Code

In the course of developing more sophisticated programs, mistakes will often leave Roboarm in some unusual position. In these cases, the ROBOARM1 program is a handy utility to have around—simply run it to restore Roboarm to its starting position.

THE TEST MAT

Roboarm is a fairly general-purpose robotic arm capable of grasping a variety of objects. For consistency, however, the remainder of this chapter assumes that the white hubs for treads are being picked up by Roboarm. Although not strictly required, a test mat for Roboarm can also be helpful. The test mat shown below indicates the position for Roboarm, the circular path traversed by the arm in its lowered position, and the point on the circle corresponding to home position. The mat should also be a dark color to help the light sensor reliably identify the white hubs. A 22" x 28" sheet of black or dark green poster board works well. The actual home position should be determined empirically. Place Roboarm in the box so that the alcove in the base is toward the top of the box. Now run the ROBOARM1 program and note the final position of the arm (which is its home position).

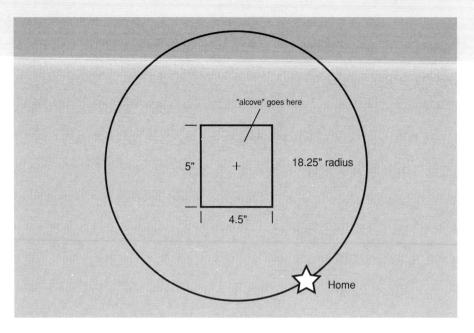

Figure 19-42:
Roboarm Test Mat

CLEANING UP

Roboarm's first real assignment will be to "clean up its room." The four white hubs will be put at various positions along the circle, and a small container will be placed at the home position. Roboarm must then pick up each of the hubs and deposit them in the container.

For each hub, Roboarm must take the following actions: find the hub, pick it up, and drop it in the container. One way to find a hub is to have Roboarm spin around until the light sensor detects a white-colored object. Then the existing **grab** and **up** operations can be used to pick up the hub. Since the container is in the home location, the **home** and **release** operations can then be used to move the arm over the container and drop the hub. The arm then needs to be lowered (preferably just past the container) in order to proceed onto the next hub. The NQC version of this program follows on the next page.

```
// roboarm2.nqc
// program to put 4 hubs away

// include the basic operations
#include "roboarm.nqh"

// threshold to detect a hub
#define THRESHOLD     38

// how far past "home" to spin
#define SPIN_DELAY    50

// number of hubs to put away
#define COUNT   4

void detect()
{
    // spin until a piece is seen
    On(SPINNER);
    until(EYE >= THRESHOLD);
    Off(SPINNER);
}

task main()
{
    // configure the sensors
    SetSensor(LEVEL, SENSOR_TOUCH);
    SetSensor(EYE, SENSOR_LIGHT);
    SetSensor(HOME, SENSOR_TOUCH);

    // for each piece...
    repeat(COUNT)
    {
        detect();
        grab();
        up();
        home();
        release();
        OnFor(SPINNER, SPIN_DELAY);
        down();
    }

    // we're done
    PlaySound(SOUND_FAST_UP);
}
```

Depending on your lighting conditions, the THRESHOLD value may need to be adjusted. Another constant, SPIN_DELAY, is used to allow the arm to move past the container before being lowered back into position to search for another hub. Depending on the size of your container, this value can be made larger or smaller.

The direction that Roboarm spins while looking for a piece is not arbitrary. The light sensor is slightly offset from the center of the grabber. Because of this, rotation in a counterclockwise direction allows the sensor to detect objects just before they are within Roboarm's grasp. If the arm were rotated in the other direction, the sensor would always be detecting objects a little too late.

A version of the program in RCX Code is shown below. Like the previous program, it makes use of the basic operations introduced earlier (**home, up, down, grab,** and **release**).

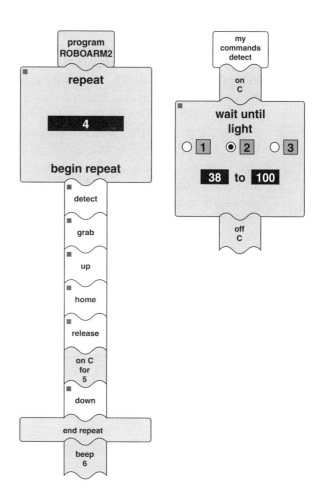

Figure 19-43:
Cleaning Up in RCX Code

Before the program is run, the arm should be level with the grabbers open. Since a container will be sitting at the home position, the arm will need to be rotated slightly past home. One way to accomplish all of this is to run the ROBOARM1 program to get the arm level and open, then manually spin it a bit counterclockwise past home. Manual adjustment of Roboarm is often easier when the RCX is turned off. If all goes well, this adjustment will only be required one time since the "clean up" program leaves the arm in a suitable position after it is completed.

LINING UP

Roboarm's second assignment is to look for the hubs and, instead of dropping them into a container, neatly line them up. In addition, rather than always expecting a fixed number of hubs, it should automatically quit once no more hubs are available. This assignment is a little too complicated for RCX Code to handle, so only an NQC version of the program will be presented.

Many elements of this program, such as the process of grabbing and lifting a hub, are similar to the previous program. However, there are a few new complications. The first issue is how to determine when no more hubs can be picked up. One strategy for dealing with this is to assume that, if the arm ever rotates all the way around to home without spotting a hub, its work is done. As a consequence, the program must be simultaneously checking for two conditions: "has the light sensor detected a hub?" and "has the arm rotated around to home?" This sounds like a good candidate for using multiple tasks.

```
task pickup()
{
    while(true)
    {
            until(EYE >= THRESHOLD);
            // pick up and arrange the hub
    }
}

task done()
{
    until(HOME==0);
    stop pickup;
}
```

Unfortunately, the pickup task may have to rotate past the home position in order to arrange the hub. If this ever happened, the done task would stop the pickup task before its work was done. This implies that some communication between the two tasks should take place to indicate when it is acceptable to stop picking things up. Such communication could take the form of a variable that was set and cleared appropriately.

```
int ok_to_stop = 1;

task pickup()
{
    while(true)
    {
        until(EYE >= THRESHOLD);
        ok_to_stop = 0;
        // pick up and arrange the hub
        ok_to_stop = 1;
    }
}

task done()
{
    while(true)
    {
        until(HOME==0);
        if (ok_to_stop==1) break;
    }
    stop pickup;
}
```

This is certainly a workable approach, but I find it simpler to abandon multiple tasks and instead use a nested condition within a single task.

```
while(HOME==1)
{
    if (EYE >= THRESHOLD)
    {
        // do whatever we want with the piece
    }
}
```

Once a hub has been detected, whatever is used to handle the hub will run to completion before another check for home occurs.

The second issue is how to line up the hubs. A good approach is to use a variable to keep track of how many hubs have already been lined up. The position for a new hub is determined by first rotating the arm to its home position, then rotating it for an amount of time proportional to the number of hubs already placed. In effect, this will space the hubs roughly evenly along the circle.

We can put this all together in the following program:

```
// roboarm3.nqc
// program to line up hubs

// include the basic operations
#include "roboarm.nqh"

// threshold to detect a hub
#define THRESHOLD    38

// how far apart to place pieces
#define SPACING 20

task main()
{
    int count = 0;

    // configure sensors
    SetSensor(LEVEL, SENSOR_TOUCH);
    SetSensor(EYE, SENSOR_LIGHT);
    SetSensor(HOME, SENSOR_TOUCH);

    // spin past "home" position
    On(SPINNER);
    until(HOME==1);

    while(HOME==1)
    {
        if (EYE >= THRESHOLD)
        {
            // piece detected—grab it
            Off(SPINNER);
            grab();
            up();
```

```
            count++;

            // rotate to next "spot"
            OnRev(SPINNER);
            until(HOME==0);
            Fwd(SPINNER);
            Wait(count * SPACING);
            Off(SPINNER);

            // put the piece down
            down();
            release();

            // rotate past the piece
            On(SPINNER);
            Wait(SPACING);
        }
    }

    // our work is done...
    Off(SPINNER);
    PlaySound(SOUND_FAST_UP);
}
```

The value for SPACING controls the distance between lined up hubs. A subtle detail is that count is incremented before it is used in the Wait statement. As a result, the first hub isn't placed at home, but is slightly past home. This has the nice side effect of ensuring that after all of the other hubs have been lined up, the first hub won't accidentally be detected and picked up before Roboarm notices it got back to home (and is thus done with its work).

VARIATIONS

Roboarm is a fairly general-purpose device that can be pro-grammed for a variety of tasks such as building a small tower or sorting blocks based on their color. Currently, its detection of home and level positions is rather crude. Either of these measurements could be improved by using a rotation sensor instead of a touch sensor. This would allow for more precise positioning in either the rotational or the vertical directions. Other options could include adapting the grabbers for specific objects.

Afterword

I hope you have enjoyed building the various robots presented in this book. More important than the robots themselves, however, are the techniques used in their construction and programming. Armed with the knowledge of gear ratios and feedback systems, you are now ready to let your imagination roam and to build some truly unique creations.

The next steps depend entirely on your own preferences. For more advanced programming you can move from RCX Code to NQC, or even from NQC to alternative firmware such as legOS. From a mechanical perspective, consider investing in some of the LEGO Technic sets. Besides providing a wealth of new parts, the models themselves usually demonstrate some interesting design techniques. Although the larger sets are rather expensive, some of them are true marvels of LEGO design.

For inspiration, turn to the real world. I got the idea for Vending Machine while using a cash card to buy a candy bar, and Bumpbot was inspired by bumper cars in an amusement park. Even nature can provide ideas–some bugs scurrying around served as the impetus for Bugbot.

Whatever creations you decide to build, never hesitate to experiment and try new ideas. LEGO is remarkably easy to put together and take apart. Take advantage of this by trying out different combinations of pieces or unusual constructions. Staring at a pile of LEGO parts while trying to imagine the optimal design for a robot is usually quite unproductive. Instead, throw something together as quickly as possible, then continue to refine it. All of the robots in this book went through several iterations before I felt they were ready for presentation.

Last of all, remember that MINDSTORMS is a toy. Have fun.

Appendix A

MINDSTORMS
Sets

This appendix lists the various retail and educational MINDSTORMS sets and accessories available from LEGO. Although the MIND-STORMS sets utilize some new elements (such as the RCX), they are really just extensions of the vast LEGO universe. This means that ordinary LEGO sets, especially those from the Technic line, provide a good means of expanding your MINDSTORMS set.

The MINDSTORMS sets are divided into two groups: retail and educational. The retail sets are intended for the average consumer and can typically be found at toy stores or purchased directly from LEGO via the LEGO Shop at Home service (United States mail order), or the LEGO World Shop (online ordering). The educational sets are geared towards classroom use and are marketed by LEGO's educational division, Dacta. In the United States, LEGO Pitsco Dacta is presently the exclusive distributor of Dacta products. They also offers several "bundles" that combine all of the materials necessary to use MINDSTORMS in the classroom.

LEGO also offers MINDSTORMS accessories such as extra sensors and motors. In general the accessories are available directly from LEGO (Shop at Home and the LEGO World Shop) or though LEGO Pitsco Dacta, but not at toy stores and other retail outlets. In some cases, the educational versions have a different part number, but they are all compatible.

The listings below contain some price estimates, but these should only be used as a rough guide. Different stores may have different prices, and the prices set by LEGO may change over time. In specific, there are presently some differences in price between the educational and retail offerings, but these may get equalized over time. The bottom line is that you should shop around a little before buying.

Although every attempt was made to make this list complete, LEGO is constantly introducing new sets and accessories, so don't be surprised if a few new sets are available by the time you read this.

RETAIL SETS

#9719 - Robotics Invention System
#9747 - Robotics Invention System 1.5

These sets are the cornerstone of the MINDSTORMS lineup. It includes everything needed to get started with MINDSTORMS, including the RCX, two touch sensors, one light sensor, two motors, and about 700 assorted LEGO pieces. An infrared transmitter and cable are also provided to allow communication between a computer and the RCX.

The bundled CD-ROM contains software that runs on under Microsoft Windows and allows RCX programs to be written in RCX Code - a simple visual programming system created by LEGO. It also includes a number of challenges, which guide the user through the creation of some simple robots. A printed Constructopedia provides ideas and hints on solving the challenges.

Either set is ideal for getting started with MINDSTORMS. The assortment of LEGO pieces is complete enough to build a reasonable number of interesting robots. In fact, many of the robots in this book can be built completely from the contents of either #9719 or #9747. With a list price of $200, the sets are also a good value considering the cost of the RCX, motors, and sensors if purchased separately.

Several changes were made from the original set (#9719) to version 1.5 (#9747). Most importantly, version 1.5 of the RCX Code software is both more powerful and easier to use than the original. The mixture of pieces included in the sets has also changed, with a few less basic bricks and a few more specialized pieces appearing in the newer set.

Current owners of #9719 can purchase the Robotics Invention System 1.5 Upgrade Kit from LEGO which consists of new versions of the software, manuals, and instructions as well as a collection of the new pieces appearing in #9747. Priced around $25, the upgrade kit is a "must have" if you program in RCX Code regularly.

For expansion, a third motor and one or two additional sensors (especially a rotation sensor) are recommended.

#9730 - RoboSports

This expansion set contains 90 LEGO pieces (including a motor), a CD-ROM with additional challenges, and a new Constructopedia. The set has a "sports" theme, with challenges such as throwing a ball through a basketball hoop. With a list price of $50, this set is a relatively poor value. Unless you really want to see the challenges and Constructopedia, you are much better off purchasing the motor separately and buying LEGO Technic sets for extra pieces.

#9732 - Extreme Creatures

This expansion set includes 149 LEGO pieces. Although no additional motors or sensors are included, the set does contain the LEGO "fiber optic" unit. The set also features a CD-ROM with additional challenges and a Constructopedia. As its name suggests, the challenges in this set have a "wildlife" theme. The "fiber optic" unit is often used as a decorative piece, but it can be coupled with a light sensor to fashion a crude rotation sensor. Some of the other included pieces are useful, but with a list price of $50, this set is a relatively poor value.

RECENT ADDITIONS

Two new sets have been recently added to the MINDSTORMS line-up: #9735 Robotics Discovery Set ($149), and #9748 Droid Developer Kit ($99). Neither of these sets uses the RCX common to most of the other MINDSTORMS sets. The Robotics Discovery Set contains a "Command Center," which looks like a blue RCX and can be programmed from the buttons on its front panel (no PC required). The Droid Developer Kit contains a "Micro Scout" which is a smaller and more limited type of programmable brick. Both sets provide less expensive ways to get started with robotics, but for maximum flexibility, sets based on the RCX (such as #9719 or #9747) are recommended.

EDUCATIONAL SETS

#9725 - Amusement Park

This set is the entry level construction set for educational use. It includes two touch sensors, one light sensor, two motors, one light brick, and nearly 300 LEGO pieces. The set does not contain an RCX, the Infrared transmitter, or any software; these must all be purchased separately. One of the strengths of the Amusement Park is the set of instructions for 4 models, all within the Amusement Park theme, that provide a framework for introducing MINDSTORMS into a classroom environment. With a list price of $100, this set is an average value.

#9780 - ROBOLAB Starter Set

This is a larger educational set with over 1600 pieces. The set does not contain an RCX, Infrared transmitter, or any software; these must all be purchased separately. It is designed to be used as four separate individual sets, each of which has its own unique challenges. A large number of resource bricks are also provided. The special elements included are: 3 motors, 1 micro motor, 3 light sensors, 4 touch sensors, and 4 light bricks. With a list price of $307, this set is an average value.

#9713 - Infrared Transmitter and Cable Pack

This is the Infrared Transmitter that allows a computer to communicate with the RCX. It is a necessity for using the RCX, but you only need one per computer. List price: $25.

#9833 - AC Adapter

This is a "wall wart" that that allows the RCX to be powered from AC power instead of batteries. This is a good option if most of your creations stand still, but for mobile robots you will want to use batteries (preferably rechargeable ones).This adapter cannot be used with the RCX included in set #9747.

ROBOLAB SOFTWARE

ROBOLAB is the software used to program the RCX in classroom settings. It runs on the Macintosh as well as under Windows, and is in many ways much more powerful than the RCX Code environment provided with the Robotics Invention System (#9719, #9747). Pricing varies depending on the actual package (CD only, CD plus user's guide, or site license). A brief description of the ROBOLAB software can be found in Appendix C.

Accessories

#9709 - RCX Programmable Brick

This is the brains of any MINDSTORMS creation. An RCX is included in set #9719 and #9747, but must be purchased separately when using educational sets #9725 or #9780. Priced between $120 and $130, this is an incredibly expensive piece of LEGO, but it is also the coolest. If you have deep pockets, owning more than one RCX opens up new possibilities with MINDSTORMS.

#9757 - Touch Sensor

All of the basic sets (#9719, #9725, and #9780) already contain at least two touch sensors. Priced around $10, this accessory consists of a touch sensor and connecting wire. A similar accessory (#9911) is available exclusively from Dacta.

#9758 - Light Sensor

All of the basic sets (#9719, #9725, and #9780) already contain at least one light sensor. Additional sensors can be purchased for about $20. The Dacta version of this sensor (#9890) has a longer wire and is priced around $30.

#9756 - Rotation Sensor

The rotation sensor, like the light sensor, comes with a built in wire. It is an extremely versatile sensor, and highly recommended. The sensor retails for about $15. The Dacta number for this sensor is #9891.

#9755 - Temperature Sensor

The temperature sensor also includes a built in wire. It is something of a special purpose sensor and priced at about $25. If you have a specific project in mind that requires a temperature sensor, then by all means get one, otherwise stick with more versatile pieces. The Dacta number for this sensor is #9889.

#9738 - Remote Control

This device is similar to a television remote control, but instead of changing channels it allows you to directly control the RCX. There are buttons to control each of the motors (in either direction) and start or stop programs. In addition, the remote can send three different "messages" to the RCX (similar to how one RCX sends a message to another). The remote control costs about $20.

#5225 - 9V Motor

This is the same motor included in the various MINDSTORMS sets. If you started with the Robotics Invention System (#9719, #9747) or Amusement Park (#9725) set a third motor is highly recommended, and at less than $20, this is much more economical than buying #9730.

Appendix B
Supplementary Parts

Many of the projects in the book can be built using only the contents of the MINDSTORMS Robotics Invention System (#9719, #9747). However, five of the chapters require some supplementary parts, and the projects in two of the advanced chapters can only be partially completed without some extra parts.

There are two basic ways to get extra parts: buy complete LEGO sets, or buy small *service packs* of special parts. In the long run, buying complete sets gives you more parts for your dollar, but the service packs allow you to buy only what you need. In addition, several parts (such as sensors) are only available as service packs. Service packs are available in the United States from LEGO Shop at Home (1-800-453-4652). Some parts (such as the motor and sensors) are also available individually from the LEGO World Shop (www.legoworldshop.com).

The supplementary parts required for various chapters are summarized on the next page, along with the service pack (or packs) needed. Since sets #9719 and #9747 include different pieces, two different tables are provided. Prices for the service packs are only approximate.

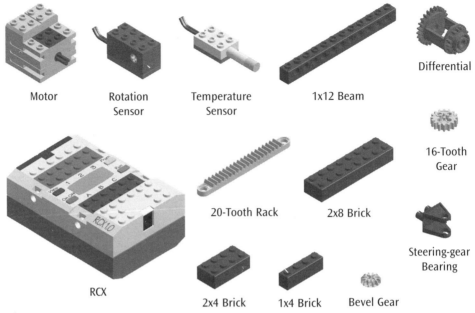

Figure B-1:
Supplementary Parts

Chapter	Parts Required (quantity)	Service Pack (price)
9 Dumpbot	Motor (1)	#5225 ($18)
10 Scanbot	Motor (1), Rotation Sensor (1)	#5225 ($18), #9756 ($15-$17)
13 Steerbot	Steering Gear Bearing (2), 20 Tooth Rack (1)	#5295 ($4)
14 Diffbot	16 Tooth Gear (2), Differential (1)	#5229 ($5)
17 Communication*	1x12 Beam (3), Motor (2), RCX (1)	see below
18 Using the Datalog*	Temperature Sensor (1)	#9755 ($25)
19 Roboarm	Motor (1)	#5225 ($25)

* These chapters may be partially completed without any extra parts.

Table B-1:
Supplementary Parts for #9719 by Chapter

Chapter	Parts Required (quantity)	Service Pack (price)
9 Dumpbot	Motor (1)	#5225 ($18)
10 Scanbot	Motor (1), Rotation Sensor (1)	#5225 ($18), #9756 ($15-$17)
13 Steerbot	Steering Gear Bearing (2), 20 Tooth Rack (1)	#5295 ($4)
14 Diffbot	16-Tooth Gear (2), Differential (1), Bevel Gear (1)	#5229 ($5)
17 Communication*	1x12 Beam (3), Motor (1), RCX (2), 2x4 Brick (1)	see below
18 Using the Datalog*	Temperature Sensor (1)	#9755 ($25)
19 Roboarm	Motor (1), 2x4 Brick (14), 2x8 Brick (1), 1x4 Brick (4)	#5225 ($25), #5144 ($6)

* These chapters may be partially completed without any extra parts.

Table B-2:
Supplementary Parts for #9747 by Chapter

The RCX-to-RCX portion of chapter 17 is intended for people with access to two MINDSTORMS Robotics Invention System sets. It is still possible to purchase the appropriate service packs, although it is not very cost effective: #5235 x 2 ($9), #5225 x 2 ($36), and #9709 ($129). At a total of $174, you'd be much better off buying a second Robotics Invention System set (#9719, #9747) and getting an additional light sensor, two touch sensors, and 700 or so other parts for about $30 more. In any event, if you don't have (or want) a second RCX, then chapter 17 can still be partially completed without any of the extra parts. Likewise, chapter 18 can be partially completed without a temperature sensor.

If you own Robotics Invention System 1.5 (#9747) and wish to complete chapter 19, then in addition to an extra motor, seven ordinary LEGO bricks will be needed. These bricks are farily common and can be obtained from practically any LEGO set. It is even possible to substitute bricks of other sizes if you don't have quite the right combination. A service pask containing these bricks can be purchased (#5114), but you will get a lot more LEGO for your money buying a "bucket" of LEGO at the local toy store.

My personal recommendation is to purchase the motor, rotation sensor, steering parts, and extra gears. This results in four service packs: #5225, #9756, #5295, and #5229, which together cost less than $45. The rotation sensor and extra motor are almost essential expansions to the Robotics Invention System set, and the extra gears and steering parts are handy to have around (Most of the time). If this sounds a bit too expensive, then consider dropping #5295 and/or #5229, but strongly consider purchasing the motor and rotation sensor.

Don't worry about a second RCX unless your pockets are pretty deep. The problem is that the RCX is only just the start, because once you have two RCXs you then need to have twice as many sensors, motors, gears, and so on. Your best bet may be to just buy another complete Robotics Invention System set. The bottom line is that there is plenty of fun to be had with a single Robotics Invention System set—don't worry about getting a second one until you've exhausted the possibilities of a single RCX.

The temperature sensor is a bit pricey and fairly special purpose. If you think you want to do a lot of experiments with it, or you have an interesting temperature-controlled robot in mind, by all means purchase one. Otherwise just read through chapter 18 without performing the temperature experiment—there are still two other datalog examples in the chapter that don't require any extra parts.

Appendix C
Programming Tools

There are a number of different tools that can be used to program the RCX. Choosing the right tool is largely a matter of personal preference. Some people prefer a graphical approach to programming, and others prefer a more traditional text-based approach. The different tools also provide varying degrees of control over the RCX, so the choice of tool also determines how powerful your programs can be. The operating system you are using also must be taken into account. Although the standard MINDSTORMS software runs only under Windows, some of the other programming tools run under other operating systems including MacOS and Linux.

RCX CODE

RCX Code at a glance:

Pros: included free with MINDSTORMS Robotics Invention System set, easy to use

Cons: limited in power, user interface can be tedious for experienced programmers

The MINDSTORMS Robotics Invention System includes software that allows users to develop RCX programs using a graphical programming system called RCX Code. This software runs under Microsoft Windows. Its graphical approach and simple block-based programming make it a good choice for those who are completely new to computer programming. Unfortunately, some of this simplicity becomes tedious for experienced users.

RCX Code's simplicity brings with it significant restrictions. Even though RCX Code relies on the same firmware as some of the other programming tools, only a fraction of the firmware's capabilities are available to the RCX Code programmer. Some of the limitations include

- Only 1 variable may be used

- The only operations permitted on the variable are "clear," "add 1," and "test"

- Only 1 timer may be used

- The datalog is not supported

The original version of RCX Code prohibited the nesting of control structures (called stack controllers), This restriction has been removed in RCX Code 1.5.

Since RCX Code is included with the MINDSTORMS Robotics Invention System, it is the default tool used to program the RCX, and many of the examples in this book are given in RCX Code. Of course, some of the more advanced examples aren't possible in RCX Code, so programs are not provided in those cases. If you're just getting started with programming, then RCX Code will probably be sufficient for your needs, and you will find plenty of material in this book to help get you going.

NQC

NQC at a glance:

Pros: powerful, fast, familiar syntax for C programmers, freely available for Windows, MacOS, Linux, and other platforms

Cons: more difficult for nonprogrammers to learn

NQC stands for "Not Quite C," which is a simple language that has a similar syntax to the C programming language. NQC was developed completely independently of LEGO, and is not affiliated with them in any way. It is more of a traditional computer programming approach; a program is written as a text file that is then compiled and downloaded to the RCX. Although this is not as easy to get started with as a graphical tool, for those even moderately familiar with programming, a textual language can be a much more efficient development tool.

NQC is free software—in both price and principle. "Free in price" means that it is available free of charge. "Free in principle" means that the source code itself is also free to anyone who wants it. Official versions exist for Windows, MacOS, and Linux. Because the source code is also freely available, unofficial ports have been made to other systems such as NetBSD, Solaris, LinuxPPC, and OS/2.

There are several wrapper programs for NQC that provide an integrated environment in which to write, compile, and download programs. For Windows, the Rcx Command Center (RcxCC) is an excellent environment. MacNQC is a similar program for MacOS users.

One final advantage of NQC is speed. Several of the other tools are awkwardly slow on anything but the fastest computer hardware. NQC runs well even on outdated hardware and can compile and download programs faster than either RCX Code or Robolab.

NQC code is provided for all of the examples in the book.

ROBOLAB

Robolab at a glance:

Pros: easy to use graphical interface, very powerful, runs under MacOS and Windows

Cons: must be purchased separately from MINDSTORMS Robotics Invention System

Robolab is another LEGO programming tool, but it is a product of Dacta (the educational branch of LEGO) and is aimed at classroom, rather than home, use. Robolab provides an intuitive graphical environment and runs under both Windows and MacOS. The Robolab environment is based on Lab View, by National Instruments, and provides a very different programming paradigm from RCX Code.

Robolab actually provides two different programming environments: one intended for simple introductory use, and the other for more advanced work. Both environments feature multiple levels that facilitate a gradual learning process by introducing new concepts in small steps. At its most advanced level, Robolab provides full access to the features of the RCX's standard firmware. Robolab also provides a full set of tools for capturing and analyzing the datalog. This makes it ideally suited for use in experimentation.

Robolab must be purchased separately from the MINDSTORMS Robotics Invention System. Because of this, Robolab examples are not given in the book, although all of the programs could easily be adapted to run under Robolab.

SPIRIT.OCX

Spirit.OCX is not a development tool itself. Rather, it is an Active X control that allows other programs, such as Microsoft Visual Basic, to construct and download RCX programs and/or directly control the RCX. Spirit.OCX is part of the MINDSTORMS Robotics Invention System software and only runs under Windows. The documentation for Spirit.OCX is available from LEGO as their Software Development Kit (SDK) for MINDSTORMS.

Programming the RCX in Visual Basic sounds pretty interesting, but Spirit.OCX doesn't deliver on this completely. Spirit.OCX does allow a Visual Basic program running on a host computer to query and control an RCX. It also allows a Visual Basic program to construct an RCX program and download it. However, the constructed RCX program is limited to using only features from Spirit.OCX. None of the familiar functions and features of Visual Basic will actually be available to the RCX program. Overall, Spirit.OCX provides some interesting functionality, but it isn't quite as powerful as one might first assume.

There are also numerous public domain and shareware tools that serve as front ends to the Spirit.OCX. These tools are too varied to cover adequately in this book.

CUSTOM FIRMWARE

Several people have created custom firmware independently of LEGO in order to enhance the RCX's capabilities. One such firmware replacement, legOS, allows the RCX to be programmed with native Hitachi H8 machine code, using traditional tools such as an assembler or C compiler. Another firmware replacement, pbFORTH, supports programming the RCX using FORTH.
Tools based on custom firmware solutions are certainly the most powerful and flexible programming tools around. The speed at which this field is growing is also a testament to the excitement that the RCX has created within the robotics community. However, detailed discussion of these alternative programming tools is beyond the scope of this book. For more information, consult some of the resources listed in Appendix E.

Appendix D
NQC Quick Reference

These pages are intended as a quick guide to the most commonly used NQC features. More detailed description of these features (along with several features not presented here) can be found in the *NQC Programmer's Guide*.

STATEMENTS

Statement	Description
while (*cond*) *body*	Execute *body* zero or more times while condition is true
do *body* while (*cond*)	Execute *body* one or more times while condition is true
until (*cond*) *body*	Execute *body* zero or more times until condition is true
break	Break out from while/do/until body
continue	Skip to next iteration of while/do/until body
repeat (*expression*) *body*	Repeat *body* a specified number of times
if (*cond*) *stmt1* if (*cond*) *stmt1* else *stmt2*	Execute *stmt1* if condition is true. Execute *stmt2* (if present) if condition is false
start *task_name*	Start the specified task

stop *task_name*	Stop the specified task
function(args)	Call a function using the supplied arguments
var = expression	Evaluate *expression* and assign to variable
var += expression	Evaluate *expression* and add to variable
var -= expression	Evaluate *expression* and subtract from variable
*var *= expression*	Evaluate *expression* and multiply into variable
var /= expression	Evaluate *expression* and divide into variable
var |= expression	Evaluate *expression* and perform bitwise OR into variable
var &= expression	Evaluate *expression* and perform bitwise AND into variable
return	Return from function to the caller
expression	Evaluate *expression*

CONDITIONS

Conditions are used within control statements to make decisions. In most cases, the condition will involve a comparison between expressions.

Condition	Meaning
true	Always true
false	Always false
expr1 == expr2	True if expressions are equal
expr1 != expr2	True if expressions are not equal
expr1 < expr2	True if *expr1* is less than *expr2*
expr1 <= expr2	True if *expr1* is less than or equal to *expr2*
expr1 > expr2	True if *expr1* is greater than *expr2*
expr1 >= expr2	True if *expr1* is greater than or equal to *expr2*
! condition	Logical negation of a condition

cond1 && cond2	Logical AND of two conditions (true if and only if both conditions are true)
cond1 || cond2	Logical OR of two conditions (true if and only if at least one of the conditions is true)

EXPRESSIONS

There are a number of different values that can be used within expressions, including constants, variables, and sensor values. Note that `SENSOR_1`, `SENSOR_2`, and `SENSOR_3` are macros that expand to `SensorValue(0)`, `SensorValue(1)`, and `SensorValue(2)` respectively.

Value	Description
number	A constant value (e.g., 123)
variable	A named variable (e.g., x)
Timer(*n*)	Value of timer *n*, where *n* is between 0 and 3
Random(*n*)	Random number between 0 and *n*
SensorValue(*n*)	Current value of sensor *n*, where *n* is between 0 and 2
Watch()	Value of system watch
Message()	Value of last received IR message

Values may be combined by using operators. Several of the operators may be used only in evaluating constant expressions, which means that their operands must be either constants or expressions involving nothing but constants. The operators are listed here in order of precedence (highest to lowest).

Operator	Description	Associativity	Restriction	Example
abs()	Absolute value	N/A	None	abs (x)
sign()	Sign of operand	N/A	None	sign (x)
++	Increment	Left	Variables only	x++ or ++x
--	Decrement	Left	Variables only	x-- or --x

-	Unary minus	Right	None	-x
~	Bitwise negation (unary)	Right	Constant only	~123
*	Multiplication	Left	None	x * y
/	Division	Left	None	x / y
%	Modulo	Left	Constant only	123 % 4
+	Addition	Left	None	x + y
-	Subtraction	Left	None	x - y
<<	Left shift	Left	Constant only	123 << 4
>>	Right shift	Left	Constant only	123 >> 4
&	Bitwise AND	Left	None	x & y
^	Bitwise XOR	Left	Constant only	123 ^ 4
|	Bitwise OR	Left	None	x | y
&&	Logical AND	Left	Constant only	123 && 4
||	Logical OR	Left	Constant only	123 || 4

RCX FUNCTIONS

Most of the functions require all arguments to be constant expressions (numbers or operations involving other constant expressions). The exceptions are functions that use a sensor as an argument and those that can use any expression. In the case of sensors, the argument should be a sensor name: SENSOR_1, SENSOR_2, or SENSOR_3. In some cases there are predefined names (e.g., SENSOR_TOUCH) for appropriate constants.

Function	Description	Example
SetSensor(*sensor, config*)	Configure a sensor	SetSensor(SENSOR_1, SENSOR_TOUCH)
SetSensorMode(*sensor, mode*)	Set sensor's mode	SetSensorMode(SENSOR_2, SENSOR_MODE_PERCENT)
SetSensorType(*sensor, type*)	Set sensor's type	SetSensorType(SENSOR_2, SENSOR_TYPE_LIGHT)

ClearSensor(*sensor*)	Clear a sensor's value	ClearSensor(SENSOR_3)
On(*outputs*)	Turn on one or more outputs	On(OUT_A + OUT_B)
Off(*outputs*)	Turn off one or more outputs	Off(OUT_C)
Float(*outputs*)	Let the outputs float	Float(OUT_B)
Fwd(*outputs*)	Set outputs to forward direction	Fwd(OUT_A)
Rev(*outputs*)	Set outputs to reverse direction	Rev(OUT_B)
Toggle(*outputs*)	Flip the direction of outputs	Toggle(OUT_C)
OnFwd(*outputs*)	Turn on in forward direction	OnFwd(OUT_A)
OnRev(*outputs*)	Turn on in reverse direction	OnRev(OUT_B)
OnFor(*outputs, time*)	Turn on for specified number of 100ths of a second. Time may be an expression.	OnFor(OUT_A, x)
SetOutput(*outputs, mode*)	Set output mode	SetOutput(OUT_A,OUT_ON)
SetDirection(*outputs, dir*)	Set output direction	SetDirection(OUT_A,OUT_FWD)
SetPower(*outputs, power*)	Set output power level (0-7)	SetPower(OUT_A, x)
	Power may be an expression	
Wait(*time*)	Wait for the specified amount of time in 100ths of a second. Time may be an expression.	Wait(x)
PlaySound(*sound*)	Play the specified sound (0-5)	PlaySound(SOUND_CLICK)
PlayTone(*freq, duration*)	Play a tone of the specified frequency for the specified amount of time (in 10ths of a second)	PlayTone(440, 5)

ClearTimer(*timer*)	Reset timer (0-3) to value 0	ClearTimer(0)
StopAllTasks()	Stop all currently running tasks	StopAllTasks()
SelectDisplay(*mode*)	Select one of 7 display modes: 0: system watch, 1-3: sensor value, 4-6: output setting. Mode may be an expression.	SelectDisplay(1)
SendMessage(*message*)	Send an IR message (1-255). Message may be an expression.	SendMessage(x)
ClearMessage()	Clear the IR message buffer	ClearMessage()
CreateDatalog(*size*)	Create a new datalog of the given size	CreateDatalog(100)
AddToDatalog(*value*)	Add a value to the datalog. The value may be an expression.	AddToDatalog(Timer(0))
SetWatch(*hours, minutes*)	Set the system watch value	SetWatch(1,30)
SetTxPower(*hi_lo*)	Set the infrared transmitter power level to low or high power	SetTxPower (TX_POWER_LO)

RCX CONSTANTS

Many of the values for RCX functions have named constants that can help make code more readable. Where possible, use a named constant rather than a raw value.

Sensor configurations for SetSensor()	`SENSOR_TOUCH, SENSOR_LIGHT, SENSOR_ROTATION, SENSOR_CELSIUS, SENSOR_FAHRENHEIT, SENSOR_PULSE, SENSOR_EDGE`
Modes for SetSensorMode()	`SENSOR_MODE_RAW, SENSOR_MODE_BOOL, SENSOR_MODE_EDGE, SENSOR_MODE_PULSE, SENSOR_MODE_PERCENT, SENSOR_MODE_CELSIUS, SENSOR_MODE_FAHRENHEIT, SENSOR_MODE_ROTATION`
Types for SetSensorType()	`SENSOR_TYPE_TOUCH, SENSOR_TYPE_TEMPERATURE, SENSOR_TYPE_LIGHT, SENSOR_TYPE_ROTATION`
Outputs for On(), Off(), etc.	`OUT_A, OUT_B, OUT_C`
Modes for SetOutput()	`OUT_ON, OUT_OFF, OUT_FLOAT`
Directions for SetDirection()	`OUT_FWD, OUT_REV, OUT_TOGGLE`
Output power for SetPower()	`OUT_LOW, OUT_HALF, OUT_FULL`
Sounds for PlaySound()	`SOUND_CLICK, SOUND_DOUBLE_BEEP, SOUND_DOWN, SOUND_UP, SOUND_LOW_BEEP, SOUND_FAST_UP`
Modes for SelectDisplay()	`DISPLAY_WATCH, DISPLAY_SENSOR_1, DISPLAY_SENSOR_2, DISPLAY_SENSOR_3, DISPLAY_OUT_A, DISPLAY_OUT_B, DISPLAY_OUT_C`
Tx power level for SetTxPower()	`TX_POWER_LO, TX_POWER_HI`

KEYWORDS

Keywords are those words reserved by the NQC compiler for the language itself. It is an error to use any of these as the names of functions, tasks, or variables.

_sensor	else	sign
abs	false	start
asm	if	stop
break	inline	sub
const	int	task
continue	repeat	true
do	return	void
		while

Appendix E
Online Resources

There is a wealth of online resources for LEGO MINDSTORMS. A brief list of a few such websites is presented below. Since the printed page moves somewhat more slowly than the Internet does, some of the links below may have changed or moved by the time you read this.

Your first stop should be this book's own website:

• **www.apress.com/mindstorms**. Additional information about the topics and robots in this book, including links to other MIND-STORMS sites.

GENERAL INFORMATION

• **www.legomindstorms.com**—the official MINDSTORMS site. On it you will find tips from LEGO's master builders, product announcements, creations from other MINDSTORMS users, and several di cussion groups. For full access you need to register on the site with a special code included in the MINDSTORMS Robotics Invention System set.

• **www.lugnet.com**—LUGNET—is a fixture in the online LEGO community, combining extensive reference material with discussion groups on a variety of LEGO-related topics. The robotics discussion group (and its subgroups) are of particular interest to MINDSTORMS users.

SHOPPING

- **www.legoworldshop.com**—the LEGO World Shop—is an online store for purchasing sets directly from LEGO. The store carries a full line of MINDSTORMS sets, including accessories such as sensors and motors.

- **www.pitsco-legodacta.com**—Pitsco LEGO Dacta—is the distributor of Dacta products within the United States. They carry the educational MINDSTORMS sets and ROBOLAB software.

PROGRAMMING TOOLS AND ADVANCED TOPICS

- **www.enteract.com/~dbaum/nqc**—the home page for NQC (Not Quite C). Check here for updates to NQC, links to NQC-based robots, and other NQC information.

- **www.cs.uu.nl/people/markov/lego**—the home page for RcxCC—an easy to use development environment based on NQC.

- **www.lego.com/dacta/robolab**—the official site for ROBOLAB—the educational version of MINDSTORMS software.

- **www.crynwr.com/lego-robotics**—a repository for information and links about the internal operation of the RCX—alternative development tools, creating custom sensors, and many other advanced topics.

- **www.noga.de/legOS**—the home page for legOS—a replacement operating system for the RCX that allows programming in assembly language, C, or other languages.

- **www.hempeldesigngroup.com/lego/pbFORTH**—the home page for pbFORTH an implementation of FORTH for the RCX.

Index

License Agreement (Single-User Products)

THIS IS A LEGAL AGREEMENT BETWEEN YOU, THE END USER, AND APRESS. BY OPENING THE SEALED DISK PACKAGE, YOU ARE AGREEING TO BE BOUND BY THE TERMS OF THIS AGREEMENT. IF YOU DO NOT AGREE TO THE TERMS OF THIS AGREEMENT, PROMPTLY RETURN THE UNOPENED DISK PACKAGE AND THE ACCOMPANYING ITEMS (INCLUDING WRITTEN MATERIALS AND BINDERS AND OTHER CONTAINERS) TO THE PLACE YOU OBTAINED THEM FOR A FULL REFUND.

APRESS SOFTWARE LICENSE

1. COPYRIGHT. The SOFTWARE copyright is owned by Apress or its suppliers and is protected by United States copyright laws and international treaty provisions.

LIMITED WARRANTY

LIMITED WARRANTY. Apress warrants that the SOFTWARE will perform substantially in accordance with the accompanying written material for a period of 90 days from the receipt. Any implied warranties on the SOFTWARE are limited to 90 days. Some states do not allow limitations on duration of an implied warranty, so the above limitation may not apply to you.

CUTOMER REMEDIES. Apress's entire liability and your exclusive remedy shall be, at Apress's option, either (a) return of the price paid or (b) repair or replacement of the SOFTWARE that does not meet Apress's Limited Warranty and which is returned to Apress with a copy of your receipt. This limited warranty is void if failure of the SOFT-WARE has resulted from accident, abuse, or misapplication. Any replacement SOFTWARE will be warranted for the remainder of the original warranty period or 30 days, whichever is longer. These remedies are not available outside of the United States of America.

NO OTHER WARRANTIES. Apress disclaims all other warranties, either express or implied, including but not limited to implied warranties of merchantability and fitness for a particular purpose, with respect to the SOFTWARE and the accompanying written materials. This limited warranty gives you specific rights. You may have others, which vary from state to state.

NO LIABILITIES FOR CONSEQUENTIAL DAMAGES. In no event shall Apress or its sup-pliers be liable for any damages whosoever (including, without limitation, damages from loss of business profits, business interruption, loss of business information, or other pecuniary loss) arising out of the use or inability to use this Apress product, even if Apress has been advised of the possibility of such damages. Because some states do not allow the exclusion or limitation of liability for consequential or incidental dam-ages, the above limitation may not apply to you.

This Agreement is governed by the laws of the State of California.

Should you have any questions concerning this Agreement, or if you wish to contact Apress for any reason, please write to Apress, 6400 Hollis Street, Suite 9, Emeryville, CA 94608.